# CHILDBIRTH, MIDWIFERY AND CONCEPTS OF TIME

# Fertility, Reproduction and Sexuality

**GENERAL EDITORS:**

*David Parkin*, The Institute of Social and Cultural Anthropology, University of Oxford.

*Soraya Tremayne*, Co-ordinating Director of the Fertility and Reproduction Studies Group and Research Associate at the Institute of Social and Cultural Anthropology, University of Oxford, and a Vice-President of the Royal Anthropological Institute.

*Marcia C. Inhorn*, William K. Lanman Jr. Professor of Anthropology and International Affairs, and Chair of the Council on Middle East Studies, Yale University.

# Childbirth, Midwifery and Concepts of Time

*Edited by*
Christine McCourt

**Berghahn Books**
New York • Oxford

Published in 2009 by

*Berghahn Books*
www.berghahnbooks.com

©2009, 2010 Christine McCourt
First paperback edition published in 2010

**Library of Congress Cataloging-in-Publication Data**
Childbirth, midwifery and concepts of time / edited by Christine
McCourt.
 p. cm.
 Includes bibliographical references and index.
 ISBN 9781845455866 (hbk) -- ISBN 9781845452940 (pbk)
1. Childbirth. 2. Midwifery. 3. Public health. 4. Time -- Sociological
aspects. I. McCourt, Christine.

 GN482.1 .C55 2009
 618.4--dc22

                                                           2009032994

**British Library Cataloguing in Publication Data**
A catalogue record for this book is available from the British Library


Printed in the United States on acid-free paper.

ISBN: 978-1-84545-586-6 (hardback)
ISBN: 978-1-84545-294-0 (paperback)

This book is dedicated to my mother, for whom time ran out during its writing

# CONTENTS

## Part III Time and Childbirth Experiences

# LIST OF FIGURES

# ACKNOWLEDGEMENTS

We would like to thank all those who contributed to the studies on which this book is based, without whom it could not have been written. We hope it does justice to their experience. We would also like to thank the friends, colleagues, families and partners who supported us during the work and the writing, who are too many to name. Thanks also to Berghahn for their excellent support and to the critical reviewers who contributed their thoughts. Finally, thanks to Ronnie Frankenberg, for his inspiration and for taking the time to read and provide a foreword to this book.

# FOREWORD

## Ronald Frankenberg

Philosophers have endlessly written about death, about living towards death, about life after death and about finitude and fortitude as experienced in the face of death. However there were very few philosophers prior to the feminist philosophers who took birth into account in the analysis that they offered of freedom, self-identity, virtue or the good life. Reading many philosophers we might, indeed, suppose that man experienced himself always first in isolation from others; that he never had to learn where the boundaries of his own self, his will and his freedom lie; and that he (or rather she) does not carry within himself (or rather herself) the gradual capacity to become two selves. We are lacking models that explain how identity might be retained whilst impregnated with otherness, and whilst other selves are generated from within the embodied self. This lack of theorization of birth – as if birth was just 'natural', something that happened before man 'is' – might be most evident in some continental philosophers (in Heidegger, for example, whose theorization starts with an existent who is simply 'thrown into the world').

—Battersby (1998: 17–18)

The introduction to this book reveals how ethnographers have, in their study of very diverse and scattered societies gone about showing the importance of time in the construction and analysis of childbirth behaviour and experience in different societies and groups across the world. My aim in this foreword, therefore, is to encourage readers to see this work in a more general and more theoretical context of how people involved in many diverse societies throughout history have successfully constructed and interpreted their reproductive experience and, especially, the extent to which different actors and genders in the process were sometimes in, but more often out of, control of their own bodily experiences in terms of both their duration in time and of their temporary and permanent outcomes. This leads me to open some discussion of creative imaginative literature alongside philosophical theories and the warring perceptions of lay men and women, doctors, nurses and male and female midwives; and to open speculation about what part all this might play in still unresolved gender and 'professional' conflicts about time in pregnancy and the practices surrounding it.

It is in one sense a very modern book on a very modernist theme (although essentially and, in respects of avoidable danger and mortality, unfortunately, an apparently eternal one). The book is based on work done within, and shared significantly between, the two disciplines of anthropology and midwifery and their shared objects of concern. It takes issue with the supposed unique nature of temporal experience and, in an important sense, the disputed ownership of the right to control the use and define the 'duration time' as period (see definitions below), which itself arises from the lack of individual women's autonomy even up to this day, concerning the establishing of socially and culturally defined gender terms in even those activities to which they have in natural terms been biologically committed. In the modernizing society of the eighteenth century in Europe and elsewhere, aspects of control were developed and imposed which created tensions that still remain unresolved. These clashes, in Britain, were already rooted in disputation between the sexes and generations, before and during that period. This has continued to be so during childbirth, especially for individual women. On the day on which I first wrote this, a Danish woman MEP and former priest (Hanne Dahl) is pictured in English newspapers voting with one hand raised while with the other she steadies her two-month old baby,[1] since there was no provision at the Strasbourg parliament either for nursery care or maternity leave for women members, let alone for their male partners! It was, it appears, about to be considered in the following week.

Although a working definition of the meaning of the expression 'time' in relation to pregnancy is difficult to find, one of the most effective in this context seemed to me to be the undated, and unsourced, one on the small cardboard box of some Accurist wristwatches, namely, 'The continuous passage of existence in which events pass from a state of potentiality in the future, through the present, to a state of finality in the past'. The *Oxford English Dictionary* presents so great a number of differing definitions of time that it defies a secure rational choice, so I fell back on a particular edition of *Webster's* in which meaning 7 seemed to me to be appropriate:

A period of duration set or thought of as set; specifically

(a) a period of existence; a life time; as 'my time is almost over.'

(b) the period of pregnancy [!]

(c) a term of apprenticeship

(d) a term of imprisonment [!][2]

An iconic figure for whom the experience is documented was the early, at that time almost unique in England, professional writer and feminist, Mary Wollstonecraft, author of the great work *Vindication of the Rights of Women* (1792), who was born in 1759 in London and died of puerperal, 'childbirth', fever in 1797. She had written earlier:

But what have women to do in society? I may be asked, but to loiter
with easy grace; surely you would not condemn them all to suckle fools,
and chronicle small beer! No. Women might certainly study the art of
healing, and be physicians as well as nurses. And midwifery, decency
seems to allot to them, though I am afraid the word midwife, in our
dictionaries, will soon give place to accoucheur [male midwife derived
by masculinizing the original feminine term in French: *accoucheuse*] and
one proof of the former delicacy of the sex be effaced from the language.
(Wollstonecraft 1792: Ch IX)

The present book collects together papers about the implications, for
women in childbirth, of time and its culturally determined management
in terms of duration, as a concept and as a constraint. It indicates the
ways in which this happens independently of the personal wishes of any
individual mother. It also concerns itself with the ongoing experience of
the women's families and wider kin, especially their reproductive partners
and their own immediate forebears. It looks at the experience and role
of others involved, like those of their professional or quasi-professional
carers, before, during and after birth. These carers are imposed on them
by the conventions of their particular society and their positions within
it. They may be incumbents of specialist roles – kinship, cultural or paid
professional. Like other books in this series, the focus in each regional
context is on the specific customs and practices, and techniques and
ideologies, which colour and determine the temporal nature of the birth
experience for all those involved in it.

    The method chosen by the various writers is, unsurprisingly,
ethnographic: attending, watching, assisting, describing, inquiring,
listening and analysing. Taking note of what happens to whom, and who
says or does what, without necessarily reaching arithmetically precise, or
statistically analysing, numbers of instances or cases. The investigators –
trained in social or cultural anthropology or midwifery, and sometimes
both – seek out, if possible, all the people directly concerned in the process
of the emergence of a new human being, observing each person involved
and directly observing and discussing the timescale and other aspects of
their behaviour. This is not a new activity in history although the presence
of self-styled specialist ethnographers as observers is relatively new. While
still obviously unusual in the majority of births, it is a feature of fairly
recent origin. I have spelt out the theoretical advantages and difficulties
of ethnography as a general scientific method elsewhere (Frankenberg
2008). There, I argue that the strength of ethnography lies precisely in its
orientation, for the investigator and for the subject, towards the future,
which is in marked contrast to that of the questionnaire and formal
interview which merely describe past or present custom. I remain totally
convinced, despite the present enthusiasm for randomized controlled
trials, formal interviews and statistical analysis, that there is still a place
in the social study of socio-biological events for direct observation of, and
interpersonal relationship with, the human subjects of study. Cochrane
himself, to whom is attributed the development of formal statistical

method as the basis of evidence-based medicine, and in whose memory the Cochrane Collaboration which promotes it worldwide was created, conceded this in his (posthumously published) autobiography. He did not himself swoop down, from outside, as some of his successors have done, abstractly to measure the realities of others' somatic fortunes and misfortunes; rather, he first paused and lived locally long enough in his chosen locality, marked as it was by poverty and industrial disease, to understand the factors producing local and regional lack of well-being and finally remained to check the outcomes of applying his findings. He also generously and fulsomely praised Tudor Hart's alternative policies of praxis (combining treatment, research and direct individual intervention), of sharing and quantifying the past and present to pose questions about the future and its present indicators. Cochrane's admirable and necessary focus on industrial dust disease among male miners led him to concentrate on their individually embodied pathology (see Cochrane and Blythe 1989:162; Tudor Hart 1988).

There have usually been people in most cultures entitled, or indeed required, to be present as witnessing observers and participants of childbirth, as well as those of both sexes prohibited to attend. In many societies, not least in those with the most complex economic and industrial development, the kind of people attending at and attending on childbirth has changed in very significant ways related to other changes in society as well as to the different strata within it. Mary Wollstonecraft was an early protagonist of considerable ability at the centre of such activity, as well as of the rights of women in general, many of whom died painfully (as she herself did unnecessarily despite her relative wealth and education) in and through unreformed ways of organizing childbirth. Mary Wollstonecraft's full life is concisely documented and referenced in *The Oxford Dictionary of National Biography* by Barbara Taylor.[3] She was in fact safely delivered of her daughter Mary but became ill on the third day following the birth since the placenta had remained behind, and the surgeon who had been summoned to remove it, which he did successfully, infected her with puerperal fever at the same time, from which she died painfully within a week, as commonly happened at that period.[4] As Burch points out, puerperal or childbirth fever was a mystery, especially because the efforts of both doctors and hospitals seemed to make it both more prevalent and more severe. Wherever the medical men went, the disease grew more common, and in their hospitals it was commonest of all. Burch also writes of Campbell, a well-known Scottish doctor in 1791. He first denied the contagiousness of puerperal fever, but personal experience changed his mind. He dissected the corpse of a woman killed by the disease, putting her uterus in his coat pocket so that he could show it to his students. He felt that neither gloves nor hand washing was needed. 'The same evening', he wrote, 'without changing my clothes, I attended the delivery of a poor woman in the Canongate; she died. Next morning I went with the same clothes to assist some of my pupils who were engaged with a woman in Bridewell, whom I delivered with forceps. She died'. Burch comments: 'Campbell's language as well as his report is a reminder that no one then spoke of delivering a baby. Obstetricians and midwives talked of delivering women – i.e. delivering them from the perils of childbirth. Cleanliness was associated with gender and class in preference to soap and water, let alone antiseptics or sterile environments'.

Unfortunately, considerations of space, time and the ignorance derived from the absence of the last, allow the inclusion of only a very brief mention here of the most sophisticated, but alas fictional, fanciful, satirical and complex if contradictorily real-seeming ethnographic account of time, childbirth and gender, Tristram Shandy, written by Sterne in 1797, and which is referred to several times in this book. One relevance of this novel was first brought to attention in chapter seven of the last published work of the late Mary Douglas, *Thinking in Circles: An Essay on Ring Composition*; 'Tristram Shandy, Testing for Ring Shape', (Douglas 2007).

Its wider possibilities of understanding have also been enhanced by the publication of a study of *Tristram Shandy* by Anne Bandry-Scubbi and Madeleine Descargues-Grant (2006). The English summary of the study reads:

> Tristram's protracted birth is the central but not the only occasion for the richness of the prevalent metaphor of delivery in Tristram Shandy, whose foregrounding of the book as physical object is a doorway into its creative processes. There are three births in the novel: that of Tristram himself, that of the text that bears him, and that of the reader, a reader in whose post-natal care Sterne had a commercial interest as well as a cultural interest. Behind the satire of the male midwife Dr Slop's intervention, the text of Tristram Shandy actually proposes an alternative model of midwifery, one which uses the interactive text to engineer the reader's willing collaboration and interpretation. The [literary] critic, who is occasionally castigated in Tristram Shandy, is really an ally of the narrator in the birth of the reader. The novel reveals itself as an example of Socratic method giving birth to independent minds and paradoxically enabling the mythical Shandy family to outlive itself outside the novel by way of generations of readers (Descargues-Grant 2006, pp 401).

The first thing anthropologists have in common with midwives especially, but also with some members of other creative and para-creative careers, is that they do not spring fully formed out of courses of study at university or other didactic institutions but rather create themselves as observers and interpreters in the process of creating their participation with, and analysis of, the other. As Descargues-Grant points out, Sterne's novel potentially and crucially gives birth to at least three outcomes in post-Shandy social relationships. There is a sense in which the 'time course' of all ethnography extends backwards and forwards from the limited time actually spent in the field which is not parallelled in more formal and allegedly more scientific 'objective' fields of social enquiry. Just as both Wollstonecraft and Sterne point out, parents are transformed into parents by the same social processes that produce their children, so ethnographic students of childbirth are born of the same general social processes.

Economists and 'scientific' sociologists (paradoxically) use social distance to achieve 'objective' results. The anthropological model of ethnographic involvement is life changing for subject and object, and as it was taught by Gluckman, as in the case of some other anthropologists, it was avowedly closer to a more than usually bilateral, even equilateral, form of (quasi-Freudian) 'socioanalysis'. It is one of the dialectical joys of English that by being subjected you become an object! In my case, living in North Wales and (very imperfectly I must confess) learning the language and its related sensitivities was a factor in choosing to teach and later research both amongst the miners of South Wales and at The University of Wales and in Italian hill country. Midwives, in their manifestation as *accoucheuses*, have after all at least the possibility of themselves having directly shared their subjects' experiences – by attending and waiting on labour and birth, or by personally being able to share a similar crucial and enduring experience of giving birth.

## Notes

1. See, e.g., *Daily Telegraph*, 28 March 2009.
2. Source: *Webster's Universal Unabridged Dictionary*. 1983. Cleveland, OH: Dorset and Baber, p.1911.
3. Retrieved on 25 March 2009 from: http://www.oxforddnb.com/view/article/10893.
4. Source: Burch, D. 2009. 'When Childbirth Was Natural and Deadly'. Retrieved on 25 March 2009 from: http://www.livescience.com/history/090110-natural-childbirth.html. All other references to Burch are to this article.

## References

Battersby, C. 1998. *The Phenomenal Woman*. Cambridge: Polity Press.
Bandry-Scubbi, A. and M. Descargues-Grant. 2006. *Tristram Shandy: Laurence Sterne*. Paris: Armand Colin for the Centre National d'Enseignement à Distance.
Cochrane, A.L. and M. Blythe. 1989. *An Autobiography of Professor Archie Cochrane*. London: British Medical Association.
Descargues-Grant, M. 2006. 'The Obstetrics of Tristram Shandy', *Etudes Anglaises* 59(4): 401–413.
Douglas, M. 2007. *Thinking in Circles*. New Haven, CT: Yale University Press.
Frankenberg, R. 2008 'The Role of Ethnographic Argument in the Prediction and/ or Creation of Social Futures', *Twenty-first Century Society* 3(2): 175–85.
Tudor Hart, J. 1988. *A New Kind of Doctor*. London: Merlin Press.
Wollstonecraft, M. 1792. *Vindication of the Rights of Woman*, Accessed 17 April 2009 at: http://womenshistory.about.com/library/etext/bl_vindication009.htm

# INTRODUCTION

This book brings together and explores writings on the theme of time in relation to childbirth. The contributors include anthropologists, and midwives who have found anthropological approaches useful in their work. We aim to present a comparative approach, in order to gain wider insights from analysis of different cultural and organizational settings. Much of the work included in the book has taken place in so-called Western[1] or biomedical[2] settings, but nonetheless involves a comparative element by including a variety of cases and attempting to learn from the differences and similarities between them. As well as cross-cultural comparison, we look at differences in concepts, experiences and approaches within Western settings, and particularly the recent development of 'alternatives' to biomedicine which have developed in response to many concerns about the medicalization of childbirth, including the ways in which time is managed.

In this book we also aim to show, through case studies, how anthropological methodology and theories have been used by maternity researchers, including practitioner researchers, to help them take a different look at the familiar world of practice and to 'make it strange', to enable a fresher or more open and critical focus to emerge. The different studies show how biomedical practices are not always evidence based, for example, but deeply rooted in established hierarchies of thinking and practice. They also illuminate the ways in which beliefs about time, and the way it is managed, are integral to biomedical practice, and found in biomedical settings such as obstetric hospitals, as much as they are in settings which are commonly thought of as 'traditional' or 'cultural'. They highlight how authoritative knowledge and practices maintain their power, as anthropologists Jordan (1993) and David-Floyd (1994) have argued, through coming to be seen as right and natural, as well as by association with professional power.

Time is a fundamental theme in considering childbirth. It is concerned with social and cultural as well as physical reproduction, and with the continuities and ruptures between generations. Childbirth forms a kind of historical moment and point of transition. Childbirth is central to all cultures, and the ways in which birth is managed are profoundly culturally shaped, so much so that it can never be described as a purely physiological or even psychological event. It is an event where different cultural assumptions, expectations, ways of doing, are 'impressed' upon the participants through the established ways of managing birth. Women and their attendants in birth are not simply passive vehicles of cultural assumptions and practices, however, but actively use and negotiate established norms (Lock and Kaufert 1998; Unnithan-Kumar 2004). The studies in this book illustrate this well, and show how women, midwives and other birth attendants are affected by issues of power and control, but also actively attempt to change established forms of thinking and practice.

The theme of time has been central in anthropological literature and theory. A number of anthropological studies have focused on time as a means of exploring and analysing the role of culture in cognition and debates about the relativism or universalism of concepts of time have formed an important thread in anthropological theory. Birth (as opposed to say death) has not been such a common theme, even though birth and death both form fundamental points of rupture or transition culturally. Nonetheless, much of the general anthropological work relating to time can be usefully applied to matters of childbirth, and Chapter 2 in this volume focuses on anthropological theory and writing both in relation to time and the increasing number of publications which look at issues of birth as well as death.

In the U.K., the twentieth century saw massive changes in the way childbirth was managed, and time, as well as place, was central to this. There has been much discussion of the shift of birth from home to hospital, and from a domestic to a public arena, but less so of the implications of the changes involved for the ways in which time is managed around childbirth. It is also an area where enormous changes in practice that took place in the 'West' were also being spread to 'non-Western' countries as authoritative knowledge and practice (Jordan 1993). The forms of measuring, marking, accounting for and managing time have played a major, but relatively unremarked, role in this. In postcolonial situations, authoritative knowledge was supported by notions of status and power associated with 'Western' medicine, and this has been particularly evident in the arena of women's health and reproduction (Unnithan Kumar 2004; Van Hollen 2003). In Chapters 2 and 7, for example, contributors discuss studies that have looked at how the development of biomedical practices was aligned with status and power in non-Western settings, so that women and practitioners found it difficult to question or challenge new birth practices or accepted them as signs of development, even when there is little evidence to support such claims. Even in wealthy and technologically developed countries such as Japan, discussed in Chapter 11, radical changes in the management of childbirth were introduced as part of rapid social changes under the influence of the U.S.-led postwar administration. As a result, biomedicine has influenced childbirth policies and practices globally, although for many women – especially poor, rural women who lack access to health care facilities – other ways of managing childbirth continue, and resistance to aspects of biomedical hegemony is also growing in a range of countries. This book explores cases in which such power and resistance may be played out through the ways in which time is conceptualized and managed.

## Anthropology of Health and Healthcare: Theory and Method

Much of the research that has taken place on health and healthcare internationally has taken place within the contexts of other disciplines besides medicine itself. These include psychology and social sciences, management studies and economics. Nonetheless, research within medicine has tended to confine itself to clinically focused research on diagnosis, aetiology and treatment of disease (Good 1994; Martin 1989).[3] Recent exceptions to this have been the interest in illness narratives, which has grown from collaboration between anthropologists and medicine (Kleinman 1988; Good 1994) and interest in complexity theory as a means of understanding the complexity of health care practice as well as disease and illness (Downe and McCourt 2008).[4] Biomedicine has tended to view its status as universal, lying outside the domain of cultural systems, so the role of social sciences has been mainly in areas such as the understanding of patients' beliefs, practices and experiences, or perhaps of practitioners' experiences and perspectives. However, sociologists and anthropologists have also conducted research on how healthcare is organized and delivered, and this research is more likely to treat biomedicine, even in its universalized forms, as socially and culturally situated.

Much of the early sociological work on health and healthcare did not start from the standpoint of viewing biomedicine itself as a cultural system. Instead, the beliefs and practices of other cultures, or 'folk' systems, were viewed in this way – as 'other' and therefore the proper objects of anthropological or sociological attention (Helman 1984).[5] Similarly, health sociologists were often employed to bring an understanding of the patients' perspective, or to analyse organizational and policy issues, with the aim of making the delivery of medicine more effective or efficient (Singer 1989; Young 1982). Where there was a focus on health belief systems or behaviours, this was primarily concerned with explaining why patients often do not comply with medical advice. The concern to identify health belief systems in order to improve compliance with treatment, and improve health education or prevention initiatives has also been a motivation for employing anthropologists to work on health related research, often expecting them to focus on the different beliefs and practices of minorities regarding health. One more recent example is the employment of anthropologists in public health oriented studies of HIV/AIDS, where they have been able to make significant and positive contributions (see, for example, Poehlman 2008), but we need to be mindful of working with assumptions that public health problems are primarily rooted in minorities' cultural beliefs and practices, rather than in structural inequalities. In the case of organizational studies, the emphasis might be more on efficiency – for example, analysing how professionals respond to protocols – and helping ensure the smooth running of healthcare systems. Singer (1989) has discussed how the type of employment of anthropologists – whether in independent academic posts or as contract researchers within multidisciplinary research teams, for example – may affect their capacity to stand outside the system and view it though a different lens. This is similarly a major challenge for researchers attempting to critically analyse their own discipline and sphere of practice, but the chapters in this book provide examples of how anthropological methods and theory can help practitioner-researchers do so.

As the anthropology of health and medicine has established a niche within the discipline (as with sociology) a more critical medical anthropology (Frankenberg 1980) has developed in which anthropological research is able to contribute to interdisciplinary work to improve health and healthcare while taking a different, theoretically informed and questioning view (Lambert and McKevitt 2002). The goal of critically engaged theory should be to understand the way in which medical science and medical practice take shape. It should look at forms of knowledge and practices and should aim to describe the ways that possibilities for change and improvement are limited and circumscribed (Singer 1989); in other words, to analyse medicine as a cultural system, that operates within a social, political and economic context, linking power and knowledge, rather than a value-free system that simply applies scientific knowledge, which in itself is treated as value free.

Critical medical anthropology has employed the Marxist concepts of persuasive power or hegemony (a dominant ideology or form of knowledge) to analyse the operation of authoritative knowledge that is important to areas such as medicine. Hegemony may be maintained by both structural power and the internalization of authoritative knowledge. It comes to be seen as part of the natural or cosmological order, and is maintained by all, not just the dominant group, so that it is less likely to be questioned except in situations of crisis or disruptive change or, even if questioned, remains difficult to challenge. In maternity this would apply to the roles of midwives in continuing the use of technologies that have come to be seen as part of 'the way things are done', even if not scientifically evaluated.

Midwives tend to view themselves as guardians of 'normal' or 'natural' childbirth. A midwifery philosophy of birth is one that sees birth as part of life, rather than as primarily a medical event. The orientation is not primarily one of risk, although midwives are trained in monitoring pregnancy and labour so that they can refer to an obstetrician if medical complications develop and intervention may be needed. The philosophy could be described as one of 'watchful waiting' (see Chapter 4, this volume) and has also been described by van Teijlingen (2005) as a social model of birth. In contrast, obstetric training is highly oriented towards management of childbirth as a risky medical event. The application of the medical model in hospitals in countries as varied as the U.K. and Brazil heralded routine uses of technological interventions originally designed to assist with complications in childbirth. Many midwives contested such routine practices, and took a questioning approach, campaigning and designing clinical studies to evaluate their effects (Tew 1995). However, many midwives have continued to use practices such as recumbent (lying flat) birth positions and electronic foetal monitoring, despite the lack of evidence to support their routine use, rather than the more traditional or low-tech approaches which are less intrusive to women's labours. Direct coercive force doesn't need to be applied in such a situation by individual obstetricians for the dominant ideas of what is right and proper to operate. In such a situation, although individuals and groups sometimes do challenge dominant knowledge and practices (Martin 1989), a number of practitioners and patients may also come to accept its norms (Davis-Floyd 1994). They may lack alternative sources of knowledge, or the self-confidence with which to question or challenge it. Additionally, although direct coercive force is not used, sanctions – such as professional exclusion or bullying when dominant knowledge and practices are questioned too openly (Hunter 2004) – pressure to conform to hospital protocols for risk management purposes, or fear of disciplinary action may severely constrain the choices of both women and professionals. Writers using an anthropological perspective – such as Kirkham (1989, 1996), in an ethnographic study of relations and behaviour on the labour ward – have observed how midwives, despite their concerns about dominance of obstetric theories, play a considerable role in maintaining this dominance. Such issues of knowledge, power and control form an important context to the practices around time discussed in this book, and will be returned to in different chapters. Theories of power and control in childbirth have also been developed in depth in the work of Martin (1989), Jordan (1993), and Davis-Floyd (1994, 2001), among others, on birth in the U.S.A., and such work is discussed further in Chapter 2 and subsequent chapters of this volume.

Although the anthropology of health has only recently been recognized as a distinct area within anthropology, a focus on health has always been at the core of anthropological studies, since it is so central to the life of communities. Traditionally, there has been much focus in anthropology on teleology – attempts to understand and explain the ultimate causes of things – including different cultures' attempts to explain the nature of suffering. In much of the early ethnographic work there was an explicit focus on ways of dealing with illness and death, perhaps because of their visibility and because of the importance of life transitions, which are treated ritually, in some way, in all human cultures. In contrast, childbirth was less visible in ethnographic texts. This might perhaps have reflected a male-gendered bias within anthropology, a tendency to pick up on those aspects of a culture which are seen as important or interesting, such as public and political rituals. However, childbirth has been traditionally treated as part of the domestic arena, rather than the public one, and one to which males often had no access. It was, quite literally, less visible. With a more reflexive and critical approach to ethnography in recent decades (Clifford and Marcus 1986), a greater focus on gender, and revived interest in areas such as kinship as a result of the response to the development of new reproductive technologies, that situation has changed. This work is discussed further in Chapter 2 of this volume.

Childbirth is, of course, quite different from illness, but the potential for health problems to develop in pregnancy and labour, and the reality of maternal and infant deaths, which remain high in many countries,[6] have placed the ways of managing childbirth firmly in the healthcare arena. In Western and non-Western countries alike, pregnancy and birth are subjects of medicine. Much of the history of childbirth, especially in the modern era in Europe, has been one of medicalization, an issue which will be explored further in Chapter 1. In other cultural contexts, where biomedicine forms part of a more plural system of healthcare, childbirth still touches on medical care, with traditional midwives, such as the *dais* of South Asia, often having status as healers or working alongside and acting with traditional and spiritual healers (Unnithan-Kumar 2004; Van Hollen 2003). Midwives are often seen as mediators between health and illness as well as between the material and spiritual world, and between life stages.

However, childbirth is also part of the wider subject of how women's health and reproduction are managed and socially situated. Anthropological work has been particularly valuable in highlighting how far women's reproduction is treated as a matter of concern to community and state and even to the postcolonial world order of relations between states. That is not a focus of this book, and has been well covered elsewhere (see, for example, Ginsburg and Rapp 1995; Unnithan-Kumar 2004), but it does occasionally show as a thread in the weave. In Chapter 7, for example, Becker discusses how the childbirth experiences of Canadian aboriginal women, and their forced removal from their own land and traditions, has to be understood within that wider context of attempted state management of social and biological reproduction. Even in the U.K. settings of other chapters, the lack of choices faced even by more privileged women is revealing of the complex relations between the micro- and macro-social levels of social interests at play.

In recent work a number of anthropologists working from a feminist perspective, or focusing on healthcare systems, have focused on the degree to which women's bodies are treated as metaphors or representations of wider issues (Ginsberg and Rapp 1995; Lock and Kaufert 1998; Martin 1989; Scheper-Hughes and Lock 1987; Van Hollen 2003). This is particularly true of the area of reproduction because in all cultures women's fertility and childbearing are subjects of social concern. Reproduction can be regarded as never simply an individual matter, or indeed a purely physiological or biological one. Women's individual reproduction also stands for the reproduction of society, even in societies marked by a strongly individualistic culture.

In Chapter 1, for example, we will see that the Church in pre- and early-modern Europe took a strong interest in midwifery long before state registration, since it was concerned to monitor the moral aspects of birth. Each infant born needed to be baptized into the Church and the transitions of life and death to be overseen by religious authority. The Church was powerful and intimately concerned with regulating the reproduction of society. In modern Europe, the state has largely taken over much of this concern with social reproduction from the Church. Much of this is very routine, such as the registration of births and deaths and the provision of state maternity services, but there are also more authoritarian ways in which governments have done so, such as pro-natalist policies, designed to encourage growth of the population of a state, and eugenic policies intended to discourage births and reproduction that is considered socially undesirable (Unnithan-Kumar 2004; Van Hollen 2003). Such policies have been identified (ironically perhaps) in contexts as contrasting as Nazi Germany in the 1930s (Fallwell 2002) and Israel in the 1950s (Stoller-Liss 2002). This theme is also echoed in Chapter 7, as the social and political background to the recent re-establishment of community birthing and Aboriginal midwifery in northern Canada.

The main methodology used by anthropological researchers is ethnography, and this has changed relatively little since it was established by anthropologists such as Malinowski in the first decades of the twentieth century. Although ethnographic research is increasingly accepted now in studies of healthcare, and some classic examples exist, it is still greeted with some discomfort and is not always accepted as being able to produce generalizable knowledge. Ideas about generalization in health research, rather than being theoretically based are often centred on the idea of statistical generalization from a sample to a wider population. However, anthropology was always intended to be a comparative discipline, where ethnographic studies are theoretically informed and do not stand alone. Knowledge is built up and challenged by developing themes and debates across a number of studies and social or cultural settings. This counters the criticism that ethnographic research, owing to its detailed, in-depth and time consuming nature, is often small scale and may well focus on relatively easily bounded settings, even if the boundaries are more ideal than real. Ethnographic research may utilize a range of research methods and include both qualitative and quantitative techniques.

Qualitative studies in healthcare have become more common and accepted in recent years, particularly when they focus on health beliefs, attitudes and experiences, or on the organization and delivery of care. However, many of these studies do not share a key feature of ethnographic work in anthropology, which is to analyse small-scale and local beliefs or practices with reference to wider systems. This linking of micro- to meso- and macro-social levels is important to support a critically engaged analysis of healthcare systems.

The studies described and referred to in this book mainly use explicitly and self-consciously ethnographic approaches, or have much in common with ethnography. They provide case studies of maternity care, but also case studies in how ethnographic and related qualitative methodologies, such as narrative approaches, can provide fresh insights into the operation of healthcare systems. These not only add to knowledge, in a theoretical sense, but can be applied to developing and improving the effectiveness and experience of health care.

## Outline of the Book

Chapter 1 commences by tracing historical shifts in the reckoning, marking and management of time that have occurred worldwide, particularly in the modern era. The main focus is on European history, since it was that history which gave rise to biomedicine as a healthcare system, and to the globalizing trends that have led its forms of healthcare and forms of marking time to be spread worldwide. The chapter discusses the social and economic changes that have arguably contributed to changing conceptualizations of time, but it also touches on the historical roots of capitalism and the spread and dominance of biomedicine as the authoritative knowledge and practice of universalized medicine.

In Chapter 2, a sideways step is taken to look at anthropological ideas and studies pertaining to time, and their potential value for studying and reconsidering medicine and various forms of healthcare, with a distance or 'estrangement' that can help us to both see what may be very familiar practices and ideas in a different way, and to ask new questions and to engage in a constructively critical analysis. The chapter discusses relevant theoretical perspectives from anthropology and other social sciences to support such a critical analysis, looking at the work of key thinkers such as Marx, Foucault, Durkheim and Bourdieu as well as specific ethnographic studies relevant to time and/or childbirth. It also reviews debates in anthropology about the formative roles of culture and biology in cognition and ways of organizing the environment, including ways of conceptualizing and marking time, which are particularly pertinent to cross-cultural and interdisciplinary analyses of health and medicine.

Chapters 3 and 4 discuss and analyse obstetric theories and practices of time management in labour. Chapter 3 also describes and critiques the work of Friedman and the concept of the 'active management' of labour. This approach to managing labour hinged on tight control of time and progress, and was tested in a Dublin hospital in the 1970s. As is often the case with theories that resonate well with their times, this approach was rapidly and uncritically adopted in European and North American hospitals and thence exported, in the form of authoritative knowledge and practice, worldwide. The tight monitoring and control of time in the active management of labour is a well-established example of protocol-based care, which is being increasingly advocated in 'evidence-based medicine' (EBM) and formalized by governmental bodies such as the U.K.'s National Institute of Clinical Excellence (known as NICE). The focus on EBM may have great value in questioning why certain practices or health interventions are used and whether they are effective, but the philosophy and methodology rarely acknowledges the situated nature of research, or the existence of protocols as social practices. This could be described as a project of rationalization which rests on notions of scientific evidence and rational decision making and action, but the analysis in Chapter 3 deconstructs this to show that scientific evidence itself may be highly contested, highly provisional and socially situated knowledge. Just as Bourdieu (1977) described the calendar as one of the most codified aspects of social existence, arguably the models of time and time management that childbirth protocols represent are social practices of a particularly codified form.

Chapter 4 then goes on to look at alternative theories and practices of monitoring the progress of labour used by a small number of independent midwives, mainly working outside hospital settings, and referring implicitly to more traditional forms of midwifery practice that involved watchful waiting and 'being with' the woman. This is also discussed in Chapter 8 in the context of rural Icelandic midwives who refer to 'sitting over' (*yfirseta*) the birth.

Chapters 5 to 7 continue the theme of contrasting models of managing time in childbirth, looking at changes and reforms developed recently in Euro-American birth settings that attempt to change the ways in which childbirth is managed. Chapter 5 looks at attempted reform of the way midwives' work is managed, to emulate a more traditional model of midwifery where the midwife works with and is accountable for the care of particular women, rather than working in a particular setting, accountable primarily to that service. For midwives, this change required a radically different time orientation, which can be characterized in some ways as either pre- or postmodern. Like the older midwives in history, and traditional midwives still working in resource-poor countries, the midwives had to learn to 'attend' the rhythms of women's pregnancies and labours, rather than the shift-system of clock time made necessary by hospital-based work. Stevens describes how this shift in time orientation led to reorientation in other aspects of their work, including a more woman- and community-centred approach.

In Brazil and Latin-America comparable reforms have been referred to as more 'humanized' as opposed to technocratic models of birth (Santos and Siebert 2001; Davis-Floyd 2001). Ironically, such reforms are taking place, albeit in a piecemeal and patchy fashion, in more resource-rich countries, while the models of obstetric management they attempt to reform are still being actively advocated and implemented in resource-poor countries as a form of development, despite their high cost and socially unequal spread, where healthy and wealthy women may be subject to routine active management of their labours while poor women lack access to sometimes life-saving healthcare. Similarly, models such as 'birth centres' as described in Chapter 6, refer to more traditional ways of attending[7] and caring for labouring women, while such knowledge and practices are being actively removed in 'developing' countries. Chapter 7, which looks at the reintroduction of midwifery and community birthing in a remote northern region of Canada, cuts across this apparent divide, set in a resource-rich country but in a situation where colonialism still has very strong and tangible bearing on the delivery of maternity services. Here, formally educated midwives in a setting where midwifery had been pushed to the social margins have begun to work with women who were dispossessed in more fundamental ways to bring back a sense of time and place in giving birth which is more appropriate to their sense of cultural safety.

Chapters 8 and 9 illustrate the value of using narrative approaches in health research. Chapter 8 sets out basic principles and values of using narrative approaches in research, and describes themes from narrative studies in Iceland and the U.K., arguing that a more 'storied' approach can better represent the nature of birth experience and birth care than more usually authoritative approaches to research. Chapter 9 illustrates this further with a case study of women's experiences of time in labour in an obstetrically-led birth setting in England, contrasting the women's embodied sense of time with the time management practices they encountered. This chapter also illustrates through narrative the experiential and embodied effects of the active management of labour reviewed in Chapter 3. In both sets of stories we hear reference to the ticking of clocks: by analogy, when the midwife's heart is described as 'not ticking', and quite literally when women described the imposed regime of actively managed labour as being 'against the clock'.

Chapters 10 and 11 shift the focus to the transition following birth, looking at issues of time and space in infant feeding and adapting to motherhood. Chapter 10 discusses the conflicts and paradoxes of the advocacy of 'demand feeding' in U.K. maternity hospitals, while chapter eleven looks at Japanese women's responses to the uncertainty of motherhood. Although Japanese maternity hospitals since 1945 have been run on a U.S.-inspired model, with highly restrictive practices relating to time and to infant care and feeding, Japanese notions of time and space support the women in living with uncertainty and adapting to their lives with a new baby to achieve a sense of being 'in tune'. In contrast, the 'demand feeding' policy in U.K. hospitals adopted as part of the UNICEF Baby Friendly Initiative, followed decades where women and babies were taught in Euro-American hospitals to adopt a rigid regime of measurement, by the clock and by other measures, of either the quantity of milk consumed or of height and weight plotted against time.

In settings that are in many ways both contrasting and similar, these chapters draw on critical theory to help to understand women's experiences. For example, Bourdieu's (1977) concept of habitus is used to explore the nature of embodied knowledge in a way that breaks down dichotomies between objectivity and subjectivity, knowledge and practice.

This book cannot, of course, cover all the work relevant to its themes, or all cultural settings, but throughout we refer to other work emerging in this field, much of which adopts a critically engaged perspective, and views women (as mothers or as maternity workers and carers) as active agents who negotiate the everyday reality of birth within its social and cultural context. We hope this book will make a contribution to the application of anthropology in the field of maternity care, and will also bring some useful material and insights to the discourse of anthropology.

## Notes

1.  The term 'Western' is used in this book to denote countries, usually European and North American, which are relatively resource-rich and industrially or post-industrially developed or complex. The social and economic changes that globalization has entailed mean that it is not a literally accurate term – countries are infinitely more varied – but it is widely used and understood, and seems preferable to terms such as 'developed' which carry particular value assumptions. The term 'Western medicine' is sometimes used to highlight the association between biomedicine and the status and power of Western countries.

2.  Biomedicine is a term used widely in anthropological literature to denote the form of medicine developed in Western countries. Although it developed in its cultural and historical context, as discussed in Chapter 1, it has universalizing tendencies: biomedical knowledge and practices are believed to be scientifically based, and so are considered to have universal applicability, rather than relevance to a particular culture. The universalizing claims of biomedicine have been analysed and critiqued in a number of anthropological studies, and case studies within this book add to that body of work.

3.  This observation is primarily based on my experience of designing, conducting, reviewing and reading research within a health setting. It has also been commented on by a number of sociologists, anthropologists and epidemiologists.

4.  Complexity theory draws on different disciplines, including mathematics, quantum physics and anthropology, to argue that systems which are inherently complex cannot be researched entirely through models based on linear or simple systems thinking (see, e.g., the writings of Bateson, cited in Chapter 2). Much of medical practice is highly complex and so cannot be understood by exclusive reliance on positivist approaches and forms of experimentation based on simple models. This has now been recognized by the U.K.'s Medical Research Council, but the randomized controlled trial, which is highly effective for testing the efficacy of specific interventions but less so for complex interventions, is still formally seen as the 'gold standard' form of evidence in U.K. health policy.

5.  One might argue, of course, that this was equally characteristic of much of early mainstream anthropological and sociological work. Evans-Pritchard's writings on Zande witchcraft and magic (cited in Chapter 2) could, for example, be viewed in this way, but that work has also become a useful source and inspiration for anthropologists who have turned an analytic eye towards Western healthcare beliefs and practices.

6.  For example, the World Health Organisation estimates that lifetime risk of maternal death in Africa in 2005 was 1 in 26, compared with a developed states average of 1 in 7300. Maternal and infant death rates tend to be highly correlated with income, within and between countries, and also with political factors such as structural adjustment or public health programmes and social factors such as women's status.

7.  The historical and linguistic roots of the word 'attending' are salient, as they refer to a central theme in traditional midwifery of 'waiting on' birth, which may be contrasted with the active management approach of modern obstetrics and much of midwifery practice in obstetrically led settings.

# References

Bourdieu, P. 1977. *Outline of a Theory of Practice*. Cambridge: Cambridge University Press.

Clifford, J. and G. Marcus (eds). 1986. *Writing Culture: The Poetics and Politics of Ethnography*. Berkeley: University of California Press.

Davis-Floyd, R. 1994. 'The Ritual of Hospital Birth in America', in J. Spradley and D. McCurdey (eds), *Conformity and Conflict: Readings in Cultural Anthropology*. New York: Harper-Collins, pp. 323–240.

———— 2001. 'The Technocratic, Humanistic and Holistic Paradigms of Childbirth', *International Journal of Gynecology and Obstetrics* 75:S5–S23.

Downe, S. and C. McCourt. 2007. 'From Being to Becoming: Reconstructing Childbirth Knowledges', in S. Downe (ed.) *Normal Childbirth: Evidence and Debate*, 2nd edn. Edinburgh: Elsevier, pp. 3–24.

Falwell, L. 2002. *Birthing by the Book: an Analysis of German Midwifery Textbooks*, Sheffield: Birthing and Bureaucracy: The History of Childbirth and Midwifery Conference, University of Sheffield.

Frankenberg, R. 1980. 'Medical Anthropology and Development: A Theoretical Perspective', *Social Science and Medicine* 14: 197–207.

Ginsburg, F. and R. Rapp (eds). 1995. *Conceiving the New World Order: The Global Politics of Reproduction*. Berkeley: University of California Press.

Good, B. 1994. *Medicine, Rationality and Experience: An Anthropological Perspective*. Cambridge: Cambridge University Press.

Helman, C. 1984. *Culture, Health and Illness: An Introduction for Health Professionals*. London: John Wright.

Hunter, B. 2004. 'Conflicting Ideologies as a Source of Emotion Work in Midwifery', *Midwifery* 20: 261–72.

Jordan, B. 1993. *Birth in Four Cultures: A Crosscultural Investigation of Childbirth in Yucatan, Holland, Sweden and the United States*. Prospect Heights, IL: Waveland Press.

Kirkham, M. 1989. 'Midwives and Information Giving During Labour', in S. Robinson and A. Thomson (eds), *Midwives, Research and Childbirth*, Volume 1. London: Chapman and Hall, pp. 117–38.

———— 1996. 'Professionalisation Past and Present: With Women or with the Powers That Be?' in D. Kroll (ed.) *Midwifery Care for the Future: Meeting the Challenge*. London: Balliére-Tindall, pp. 164–201.

Kleinman, A. 1988. *The Illness Narratives: Suffering, Healing and the Human Condition*. New York: Basic Books.

Lambert, H. and C. McKevitt. 2002. 'Anthropology in Health Research: From

Qualitative Methods to Multidisciplinarity', *British Medical Journal* 325: 210–13.

Lock, M. and P. Kaufert (eds). 1998. *Pragmatic Women and Body Politics*. Cambridge: Cambridge University Press.

Martin, E. 1989. *The Woman in the Body*. Milton Keynes: Open University Press.

Poehlman, J. 2008 'Community Participatory Research in HIV/AIDS Prevention: An Exploration of Participation and Consensus', *Anthropology In Action* 15(1): 22–34.

Santos, O. and E. Siebert. 2001. 'The Humanization of Birth Experience at the University of Santa Catarina Maternity Hospital', *International Journal of Gynecology and Obstetrics* 75: S73–S79.

Scheper-Hughes, N. and M. Lock. 1987. 'The Mindful Body: A Prolegomenon to Future Work in Medical Anthropology', *Medical Anthropology Quarterly* 1: 6–41.

Singer, M. 1989. 'The Coming of Age of Critical Medical Anthropology', *Social Science and Medicine* 28(11): 1193–203.

Stoller-Liss, S. 2002. *'One foot on the Steering Wheel – the Other on the Window': Israeli Nurses and Road-Delivery in the 1950s*, Sheffield: Birthing and Bureaucracy: The History of Childbirth and Midwifery Conference, University of Sheffield.

Tew, M. 1995. *Safer Childbirth? A Critical History of Maternity Care*. London: Chapman and Hall.

Unnithan-Kumar, M. (ed.) 2004. *Reproductive Agency, Medicine and the State: Cultural Transformations in Childbearing*. Oxford: Berghahn.

Van Hollen, C. 2003. *Birth on the Threshold: Childbirth and Modernity in South India*. Berkeley: University of California Press.

van Teijlingen, E. 2005. 'A Critical Analysis of the Medical Model as Used in the Study of Pregnancy and Childbirth', *Sociological Research Online*, Volume 10, Issue 2. (http://www.medicalsociologyonline.org)

Young, A. 1982. 'The Anthropologies of Illness and Sickness', *Annual Review of Anthropology* 11: 257–85.

# Part I

HISTORICAL AND CULTURAL CONTEXT

## CHAPTER 1

# FROM TRADITION TO MODERNITY: TIME AND CHILDBIRTH IN HISTORICAL PERSPECTIVE

*Christine McCourt and Fiona Dykes*

I was begot in the night, betwixt the first *Sunday* and the first *Monday* in the month of *March*, in the year of our Lord one thousand seven hundred and eighteen. I'm positive I was.– But how I came to be so very particular in my account of a thing which happened before I was born, is owing to another small anecdote... As a small specimen of this extreme exactness of his, to which in truth he was a slave,– he had made it a rule for many years of his life,– on the first *Sunday night* of every month throughout the whole year,– as certain as ever the *Sunday night* came,– to wind up a large house clock which we had standing upon the back-stairs head ... he had likewise brought some other little family concernments to the same period, in order, as he would often say to my uncle *Toby*, to get them out of the way at one time...

— Laurence Sterne, *Tristram Shandy*[1]

Delivered five kids and still looking great. Long, uncomfortable deliveries are a thing of the past.

— Advert for the Chrysler Voyager[2]

This chapter traces key aspects of the historical shift from traditional to modern and then postmodern concepts of time. The two quotes with which it begins were chosen to illustrate how this shift emerged and is reflected in aspects of popular culture. Sterne's *Tristram Shandy* has been regarded by many literary analysts as one of the first modern novels. The quote highlights the degree to which the plot and preoccupations of the story are centred on changes in notions of time of the author's day, as well as changing notions of how childbirth should be managed. The deliberate word play of the second quote, from a recent U.K. newspaper advert, associating childbirth with speed and technology, illustrates the degree to which modern, or postmodern, citizens are encouraged, indeed exhorted, to expect and demand instant results, what writers have referred to as a sense of acceleration (McGuire 1998) or the 'vertigo' of late modernity (Young 2007).

The historical shifts which we outline in this chapter took place mainly in European societies, from the early modern period onwards, but with colonialism and globalization their impact has been worldwide.[3] We then focus upon historical shifts in ways of dealing with childbirth that occurred primarily in Europe in the same time period as part of a wider social and economic history, but were then spread globally via colonialism, capitalism and postcolonial development policies. In many ways, we suggest that the changes surrounding childbirth were a microcosm of the immense social and cultural changes of this period. The changing concepts of time resonate with both.

By drawing out the parallels between the changes in the conceptualization of time and childbirth we explore the links between macro- and micro-levels of analysis that will continue to be made throughout the book. The history of maternity care parallels that of time in interesting ways. Although the focus here is primarily on European history, the implications are global, since European institutions, including biomedicine have had a far broader impact and the concepts of time discussed here have been spread via the international development agenda, in the case of health, as well as through the wider economic forces of globalization. Childbirth is a particularly good case through which to illustrate such major changes in the social world.

Our story begins with the shift from pre- to early modern Europe, highlighting very briefly the agricultural, social and economic changes which prefigured the development of industrialization and capitalism in Europe. We then focus upon the development of different modes of production, leading to the factory and the production line, which have dominated so much of modern history. Following a number of writers (most notably Thompson 1967), we suggest that the development of clock time was a crucial mechanism as well as a symbol of change. It enabled major developments within industrial societies to take place but was also, arguably, called into being and required by them. We also note the important cultural and social changes, including major changes in ways of viewing the world, which formed part of this historical movement.

## From Pre-modern to Early Modern Europe

While European economic and settlement patterns remained largely agricultural and peasant based, before the modern period, time was oriented mainly around the cyclical patterns of seasons and days (Le Goff 1986). The social historian E.P. Thompson (1967) described the work patterns of agricultural communities as primarily task-orientated, in the sense that work rhythms and use of time attended on what needed to be done, and this was closely tied to seasonal (what we shall call bio-geographical) time. For example, sheep must be attended at lambing time; crops must be gathered before the dry weather ends. The hours of the day are traced by the movements of the sun, the rhythms of plants and animals, the ebb and flow of tides.

Thompson was in effect pointing to a radical contrast between the patterns of pre-industrial work and the fragmented character of labour under capitalist development, where the tasks to be attended to are broken up and made to fit a factory production-line system. The tasks performed may not respond to the worker's perception of what needs to be done, or when, and they are often disconnected from the wider process of which they form a part. This very different form of task orientation is referred to in many of the critiques of modern healthcare provision which we touch on in the chapters to follow, where professionals' roles have changed from a focus on the whole case to performing fragmented, often repetitive tasks for a number of patients. Additionally, professionals, particularly those lower in the hierarchical chain of provision (often called the semi-professions), become infinitely substitutable for each other in regimes of time and action that are tightly monitored and governed by protocols, The historian Kahn (1989) used the term 'need–orientated' to refer to cyclical time in contrast to linear or clock time. Cyclical time, as Kahn argued, is a bodily, rhythmic time that relates to the 'organic cycle of life' in which one is 'living within the cycle of one's own body'; it is a time that is 'cyclical like the seasons, or the gyre-like motion of the generations' (Kahn 1989: 21).

In medieval Europe (circa AD 1100–1500), the pattern of landholding and use was primarily feudal and many people worked the land without a commodified form of ownership (in which land can be bought and sold like a product), a characteristic of many cultures until the more recent impact of economic globalization. Throughout this period, but particularly from around the fourteenth century, an interplay of developments led to fundamental social changes (Le Goff 1986). Among these were the movement to land enclosure, effectively creating private ownership of land and spurring the development of new agricultural technologies (Mathias 1983; Thompson 1973), and the demographic impact of the plague, and subsequent population recovery, which led to greater labour mobility (Hale 1994; Pawson 1979). Continuing developments in the specialization of craft skills and trade encouraged further urban development and the rise of a non-agricultural class in the population. The development of agricultural technologies and international trade links were also crucial triggers for the later development of secondary production technologies, particularly textiles, which required a shift from the more domestic bases of craft production to the factory (Hale 1994; Mathias 1983; Thompson 1968).

Historians of differing perspectives continue to debate the relative importance of economic, social and other factors in the changes which prefigured the rise of urbanization and industrialization in Europe, but it is clear that some interplay of the material and the social were involved in these profound changes. And as society and ways of living changed, so too did ways of viewing the world. Weber (1930) argued that cultural factors such as Protestantism were central to the rise of capitalism and the industrialization which accompanied it, while a Marxist perspective would argue for the fundamental importance of the means of production to such ideologies (Marx 1970). From this Marxist perspective time is seen in terms of exchange for money, shift work and exploitative forms of commodity production (see also Chapter 2).

Early modern Europe saw both profound changes in ways of interacting with the environment and in ways of viewing and understanding the world. The Enlightenment period in Europe can be seen as closely intertwined with the socio-economic developments we have sketched out. New concentrations of wealth no doubt facilitated the rise of an educated class with time and leisure to specialize in and pursue interests which went beyond the development of technology (Hale 1994; Hampson 1968). The work of scientists such as Galileo contributed directly to technological development, including that of the pendulum clock (Stengers 1997), while they also challenged fundamental world-views and cosmologies. It has, of course, also been argued that such changing world-views were themselves called into possibility by changing socio-economic conditions, and conditions of production (Thompson 1967). It was in this multifaceted context that the conditions of *re*production also fundamentally changed.

## The Development of the Clock and Clock Time

The development of the clock forms a point of connection between politico-economic and sociocultural changes in early modern Europe. In his seminal essay on the development of capitalism and concepts of time, Thompson (1967) described the role of clock time as central to the social as well as economic changes taking place. The requirement to standardize time and mark it in particular ways was closely related to the shift from primarily agricultural to industrial labour (Le Goff 1986).

Once labour became primarily commodified, detached from the land and domestic spaces, and as industrial patterns of work developed, time needed to be regulated in different ways and was increasingly seen as a commodity itself – something to be spent, rather than passed. Similarly, the extension of trade and transport required the scheduling of time over long distances and between different systems – timetabling.

Thompson (1967) referred to Laurence Sterne's eighteenth century satirical novel *Tristram Shandy* (quoted above) to illustrate the degree of change in habits that had taken place. Tristram is able to date his conception precisely because of Mr Shandy's habit of winding the clock at a particular time every month (Sterne 1978[1759]). Interestingly, this novel, with its focus on the tug of war over Tristram's birth and whether it should be attended by the 'old midwife' or Dr. Slop (the new style man-midwife with his instruments), juxtaposed the story of the clock's association with his conception and the conflict between tradition and modernity in childbirth which was underway at the time.

The technology of clock making developed particularly in the seventeenth century with the use of the pendulum and was regarded as an important craft. Early clocks were linked to astronomical time since they relied on ways of representing and marking the course of the sun, effectively to define the course of a day. Given seasonal and geographical variations, the speed of an astronomical clock is also variable, as the lengths of daylight hours change. The philosopher of science Stengers (1997) noted that hours of an equal length throughout the year were adopted in fifteenth century Europe, but clocks were not autonomous since the sun would be periodically in advance of, or behind, the clock. A system was adopted in Geneva in 1780 where the marking of midday was averaged throughout the year, and uncoupled from the sun's zenith, effectively meaning that social time cut through solar time. Following this, local differences in astronomical time were also cut through by social time, in order that timetables could be developed and synchronized, to facilitate railway and other forms of wider movement of goods and people. In 1845, Belgian time was organized into local zones, and in 1882 the 'unique legal time' of the Greenwich meridian was adopted (Stengers 1997). This was clearly a response to colonial expansion and the growing internationalization of capital – goods and labour – and the spread of transport and communication networks (Hobsbawm 1968). Stengers pointed out, however, that 'if the railroad still corresponds to a process of negotiation with nature and with the social, Greenwich time, from the moment it is proclaimed, is autonomised from nature and the social' (1997: 182).

Prior to the development of the pendulum clock, in 1658, the foliot clock (which relied on a complex system of falling weights) was still tied to astronomical time in that it needed to be adjusted, and could be adjusted, to daily variation in the length of hours. The pendulum clock enabled a fixed standard of time to be constructed, appearing to produce an autonomous law of time. The unit of measurement of time now became the 'second', which is calibrated according to the relationship between the length of the pendulum and the duration of its swing.[4] This technical development therefore depended on the work of Galileo in physics, which formed a major aspect of Enlightenment changes in ways of viewing the world in early modern Europe. Stengers (1997) noted the degree to which the metaphor of the clock took hold in seventeenth and eighteenth century Europe, giving the image of God as the perfect clockmaker and eventually reflected in Newton's physics, with the notion of the clockwork universe.

Thompson (1967) noted that the use of clocks was initially on a communal level – the church clock or bell, for example, marked the beginning and end of the day, as well as the key rituals and points of transition of lives and of the religious calendar (cf. Le Goff 1986). This fitted their developing role in regulating the daily and work patterns of the population and clocks were subsequently centrally placed in workshops for similar purposes. Ownership of a clock stood as a sign of social status and, as Cipolla argued (1967), the seventeenth century marked the era when growing numbers of relatively wealthy urban dwellers, for example merchants, lawyers and doctors, could afford watches and clocks, which were being made at increasingly low costs. People increasingly timed activities they would never have thought of timing and saw punctuality and time keeping as virtuous. Clocks thus not only reflected but, in turn, influenced ways of thinking as they gradually replaced the cyclical and variable times associated with the seasons with a measured time that overrode these.

The development of the more individual 'pocket watch' followed in the eighteenth century. The nature of work was changing gradually from one where a range of activities would be managed, according to need, weather and season, and the divisions between everyday life and work were less rigidly marked. Thompson (1967) argued that the typical pattern of the craftsman or pre-industrial worker was one of alternate bouts of intense labour and of idleness, a pattern that still persists in modernity only in certain self-employed occupations such as writer, artist or small farmer. We shall see in Chapter 5 how recent changes in the work of midwives, such as the model of caseload midwifery, have reintroduced such characteristically pre-industrial (or perhaps post-industrial) forms of work, where midwives' work patterns follow the needs of women on their caseload, including the unpredictable timing of labour, rather than midwives working in a shift system, caring for a range of women for short bursts of time, attending to fragmented parts of their pregnancy, birth or postnatal care. This chapter highlights the conflicts and tensions this produced for the midwives themselves, who were previously accustomed to working within a rigid shift system, and for the professionals with whom they worked within the context of the hospital system that was fundamentally unaltered.

Thompson (1967) traced the development of time first as a form of discipline, as illustrated by the use of the clock to discipline labour in the mills, potteries and textile workshops, and later employed in the development of the factory system. The system of 'clocking in' for example, commenced, despite workers' resistance, in Wedgwood's pottery in England in the eighteenth century. Second, once disciplined, time could now be calculated and valued as a resource or commodity. Here, Thompson highlighted the importance of changing world-views, as workers themselves internalized the system of time–work discipline and learnt to value and bargain with their time in a similar way: workers' resistance began to shift from fighting *against* time, to fighting about time (1967: 85). Such changing norms would also have helped to limit the degree and impact of the popular resistance to the wider socio-economic changes underway. He also referred to the part the development of state education played in attempting to inculcate such a sense of time–work discipline in the population from an early age. Children were taught to follow rigid time regimes at school, enforced by discipline until they became habituated to it, as well as taught to value time in certain ways (1967: 84). In a similar vein, Weber's arguments about the role of Protestantism in the development of time–work disciplines were neatly illustrated by reference to eighteenth and nineteenth century tracts which emphasized the immorality of idleness: time as a wasted resource (Weber 1930).

Similarly, excerpts from diaries at the time illustrate the degree to which concepts of time had shifted to those where it is the spending of time, as marked by the clock, which matters over and above the job done (Thompson 1967). This form of thinking and acting, which became an important characteristic of U.S. culture, further influenced by the 'pioneering' disposition of protestant migrants from Europe, became the precursor to the production line system as epitomized by the Ford motor factory in the twentieth century. Thompson also noted how commodified concepts of time emerged in the discourse of economic rationalism, and in the international development agenda in the twentieth century, a discourse which seems not to recognize or acknowledge its own cultural basis.

Thompson saw the socially shaped forces of economic development as inculcating new habits and ways of understanding, counting and talking about time. Similarly, Stengers (1997) argued that, with the definition of the 'second' as a unit constitutive of time, time moved beyond being a measurement to being a norm of phenomena, appearing to exist by and for itself. Such basic norms become internalized as normalized conduct. For example, as clocks were introduced across Brussels an infrastructure of habit was also established: first a town clock proclaimed the laws of time, then each person became individually responsible for knowing what the time was. So, the changes she recounted involved both a technical-economic infrastructure and social formations.

In her account of the development of clock time in Belgium, and the adoption of the internationally zoned categorization of time during the nineteenth century, Stengers (1997) argued that it is important to consider the metaphors at play in the arguments of the day about clock time. Opponents and critics of the adoption of the Greenwich meridian, to facilitate the timetabling of the railways, argued that this would disrupt the customs of the country, particularly the rural populations who still regulated their work by solar time. Indeed, Stengers quoted the burgomaster of Brussels' statement that:

> By transporting the bases of the measurement of time into the domain of abstraction, one will perhaps have gained control over the procedure used but its character will be changed. It will no longer be the natural expression of a lived reality, but will become an arithmetic or administrative expedient foreign to the most general public interests. (quoted in Stengers 1997: 181)

Stengers suggested that 'the situation of rupture is presented as a confrontation between scientific rationality and the lived world' (1997:

181) and cautioned that the opposition between science and nature which is presented is one which may lead us to overlook the degree to which such oppositions are played out in a social field. The metaphors of science and of nature form a powerful set of oppositions, which may exclude social arguments, because they appear autonomous in themselves. Stengers argued that with such metaphorical oppositions, 'social time is rendered equal with a residual part of the natural and social phenomena that have actually contributed to its development' (1997: 182) while scientific time is also seen as separated from the social process. The commodification of time, she suggested, is thereby presented as though it is part of a scientific process, and legitimized by this.

In later chapters we describe cases that illustrate her historical and philosophical argument in current contexts. In modern labour and birth, the monitoring, marking and 'active management' of time is presented as though it is simply a product of a technical process, responding to human need. The nature of the techno-scientific process appears autonomous and developments are treated as though they follow an internal logic of scientific progress quite independently of the social context (see Chapters 3 and 4, this volume). In Stengers' terms, socio-technical changes are rendered as though they are simply the products of a rationality that is external to the social, rather than a complex interplay between the techno-scientific and the social (1997: 182).

Cipolla (1967) provided a political and critical perspective on the development of the clock, contextualizing it within the socio-historical context of the economic and technological processes by which Europe gained power in the world. He linked the clock's growing valuation with the growth of industrialization and related technical machinery and with the philosophy of empiricism and utilitarianism that 'infected' all branches of human knowledge (1967: 33). The desire for power in Europe that was manifested in the combination of wars, industrialization, later capitalism and the growth of empirical science, paved the way for a central and growing place for the mechanical clock. Cipolla noted that many of the early clock makers were also gun founders and he asserted that this connection had major significance. 'The simultaneous appearance of the gun and the mechanical clock was both a testimony to the character of European development and a forecast of things to come' (1967: 40).

Cipolla argued that it was during the seventeenth century, when the scientific revolution 'exploded', that scientists saw the clock as 'the machine *par excellence*' (1967: 57). Their growing interest in the clock led to its rapid technological sophistication. Cipolla referred to these clocks as machines which like other new machines create new needs and therefore 'breed' newer machines (1967: 105). Each new tool then influences us deeply while we are using it. 'The fascination exerted by the machine induces a rapidly growing number of people into a tragic fetishism of the machine' (1967: 106–7). In the foreword to Cipolla's text, Ollard summarized what he saw as the tyrannical connections between the clock and power:

> Clocks are the prototypes for all precision instruments: and once they are valued as such and not simply admired as the most delicate and enchanting of mechanical toys the age of industrial innocence is over …. If wrist watches and guided missiles are not obtainable at one and the same shop they still to the reflective eye disclose a recognisable cousinhood. (Cipolla 1967: fwd)

Both Stengers (1997) and Thompson (1967) focused on the historical issue of the development of salaried work and how it came to be seen by many modern Europeans as a natural, human state: 'The World-as-Clock is a world in which everything works, in which the activity of each of its elements is conceived of as homogenous to the law of work' (Stengers 1997: 192). This is an important consideration when looking at medicine and at maternity care where, as with technological and 'scientific' development in general, change is presented as an inevitable and self-sustaining process of development and progress. Science is seen as having a logic and a momentum which is uncoupled from the social, and so something which cannot be questioned or changed. Conversely, there may be a tendency in critiques of medicalization to view nature as similarly detached from social processes, a kind of absolute, or even an idealization of the past, or of other cultures in the present, which are set in opposition to science. Both ideal types tend to discourage analysis of the social processes involved in historical change.

Likewise, the factory system as a model for the hospital presents itself as inevitable and necessary to the smooth flow of the system, the regular flow of work, of labour and delivery. Models of delivering healthcare are given authoritative status both by recourse to the idea of scientific rationality and to the idea of naturalness – and thus the unquestionably right way of doing things.

## Midwifery and Childbirth:
## From Traditionalism to Modernity

There are few cultures where childbirth is not a social act. Historically and cross culturally, women have traditionally given birth with the company and support of other women (Vincent-Priya 1992; Wilson 1995). Many of those women would be neighbours or kin, but midwifery as a specific occupation of women has a very long history. In most cultures, women have adopted this role, with varying degrees of specialization and learning through apprenticeship, before formal education and regulation were developed. References to midwifery can be found in early texts, including the Old Testament of the Bible.

The history of midwifery in Europe, and subsequently in North America, is part of the wider movement to modernity we have sketched out in the previous section. Not surprisingly, then, striking parallels can be drawn between historical changes in childbirth beliefs and practices, and the wider changes occurring in their macro-social context, so much so, that childbirth (or ways of entering and leaving the world of social personhood), being at the core of culture and a key point of transition, can often be used to typify the wider changes taking place in a culture. This was illustrated by Thompson's reference, in an essay explicitly focused on industrialization and time–work discipline, to the story of Tristram Shandy's conception and birth (Thompson 1967). And just as socio-economic changes in Europe were spread globally through colonialism and capitalism, the changes in childbirth that took place in Europe in the modern era have been spread worldwide, as a form of authoritative knowledge and through institutionalized practices.

Donner (2004), for example, looking at preferences for technological birth among middle class women in Calcutta, described how the first maternity wards were established in Calcutta and Madras in the second half of the nineteenth century, during British rule, although their impact was limited at the time by cultural attitudes regarding chastity and seclusion, and only became widespread from the 1950s when they were associated positively with modernity. Similarly, Van Hollen (2003) described how, in nineteenth century India, negative stereotypes of traditional midwives were circulated by the colonial administration, and the solution to the 'problem' of the *dai* was seen as promoting Western medical care to women during childbirth. This paralleled attacks on midwifery taking place in Britain at that time, a time when the safety record of hospital and medically atttended birth was particularly poor (Loudon 1992). However, while a shift was taking place from female to more male attendance in childbirth in Britain, because purdah and the caste system were assumed to be barriers to acceptance of hospital birth in India, middle and upper class British women were encouraged to undergo medical and midwifery training to volunteer in India, in the manner of missionaries: 'By penetrating this space and bringing to it knowledge of Western medicine and sanitation, not only the private space but the entire nation could become *enlightened*. In short, embedded in the discourse on childbirth was the notion that the hope for progress in the nation lay in the minds and bodies of India's women' (Van Hollen 2003: 49).

Key points of change were the movement of childbirth from the domestic sphere to the public sphere (just as the world of work moved from land and home, with agriculture and craft-based or small scale manufacturing) and from a mainly female to male sphere of authority. The place of birth moved, over time, from home to hospital, a process continuing throughout the twentieth century, and the management of birth moved from female companions and midwives to a professionalized occupation, divided between medicine and regulated midwifery. Even in remote and rural areas and in less industrialized countries, where not all women have practical access to medical facilities, the 'norm' is increasingly seen as medicalized birth, in hospitals, because of the status this signifies. This can be found in development studies and even in anthropological texts. For example, in a recent volume on reproduction, women's agency and the state, Kilaru et al. (2004) used the proportion of institutional births as a measure of progress, despite the lack of evidence that in-hospital birth is safer for mothers or infants

The historian Wilson (1995) referred to the female world of childbirth in pre- and early modern Europe. Birth normally took place in the home, but secluded from everyday activities, accompanied by a midwife and other women. This was followed by a period of confinement in which the woman rested and received care from other women. In Britain, as in most cultures, the perinatal period was marked by seclusion, and mothers normally stayed in the birth room or 'bedchamber' for about forty days, cared for by female relatives or neighbours, until their ritual reincorporation into the ordinary social world (Newell 2004). Such periods of seclusion have been traditionally found across a wide range of cultures (see Chapter 2, this volume).

Midwifery had long been a recognized female occupation, in most cultures, and the level of formality of this varied across different countries and even within them (Van Hollen 2003). The licensing of midwives by the Christian church was common in Europe from about the fifteenth century. In Paris, a system of municipal licensing of midwives was in place from 1560, and in Spain from 1523. In New York municipal 'swearing in' of midwives was introduced in 1716. Midwives varied from those who worked on an informal, part-time basis for neighbours, and were recognized within their community as a woman who helped, to those who underwent long apprenticeships with older midwives and worked on a more established basis, though usually within a specific community. Midwives' patterns of work were, therefore, very varied, and most worked according to the rhythms and needs of the communities they attended. They were 'the woman you called for' to attend births, and they might also provide care in the transition period following the birth (Leap and Hunter 1993).

From around the seventeenth century across Europe – the early modern period – men began to be more involved with childbirth, initially as surgeons who were called for in dealing with obstructed labour where the midwife could not use less destructive means to turn the baby or ease its birth. The development of forceps by the Chamberlen family allowed 'men-midwives' to intervene in a way which might save the life of the child and this technical development, as well as developments in the social status of medicine occurring more generally at the time, made their use more acceptable (Carter and Duriez 1986; Loudon 1992; Versluysen 1981). Gradually, the (male) presence of a doctor became normalized, rather than simply associated with fear and death. Wilson (1995) discussed the influences on this profound shift, noting the importance of wider social changes in Europe at the time, through which medicine came to represent both progress (to modernity) and higher social status.

The development of hospitals as a place to give birth in nineteenth century Europe went hand in hand with the development of obstetric medicine. Some historians have argued that hospitals were promoted because they facilitated the developing status of medicine as a profession, the institution and regulation of the older discipline of nursing as a subsidiary, semi-professional role, and the formal training of students necessary to a profession. Initially, hospitals catered mainly for the destitute – lying in hospitals grew out of workhouse functions – since wealthy women could retain the services of a doctor to wait on their labour at home, often alongside a midwife. This level of control of time and space was not available to most women.

As the authority of midwives was undermined in both popular culture – for example, Dickens's Martin Chuzzlewit (Dickens 1999[1843]) – and in medical texts, and as birth began to take place increasingly in hospitals, midwives themselves began to lobby and organize for more formalized systems of training and regulation that they saw as necessary to maintain status and compete with obstetrics. Led by more highly educated, urban and middle class women, there was also a concern to remove the class and gender-based derogatory images being used to undermine midwifery, which in North America saw the almost total elimination of midwifery by the turn of the twentieth century (Loudon 1992). The ensuing system of training and regulation, which culminated in the Midwives Act of 1902 in the U.K., with similar legislation elsewhere, was one which institutionalized a division of labour between obstetrics and midwifery (as dealing with abnormal or normal pregnancy and birth) and a disciplinary system of regulation which eliminated informal midwives (such as handywomen) and gave medicine ultimate disciplinary control.

This story is one which can be written as one of gender-based struggle between midwives and medical men (Donnison 1988; Towler and Bramall 1986), and as both gender and class-based struggle (Heagerty 1997). However, it also forms part of the wider story of historical change in Europe and Europe's colonial territories on which we focus here. The changes in who attended childbirth, in what place, and in what manner echo those wider social and economic changes in concepts and uses of time. For the traditional midwife, work was need orientated, in that it was entirely centred on the patterns of women's pregnancy and birth. The use of the word 'attended' to describe the midwife's role is apposite, since her use of time was different. She waited on labour. By the twentieth century we see pregnant women reconstituted as 'patients'.

Foucault's analysis (1976, 1980) emphasized the interplay of power and knowledge in this development. Siting medicine in the clinic or hospital enabled the medical gaze to be developed, and enabled the patient to be defined and constituted in certain ways. He argued that, in the shift from traditionalism to modernity, discipline changed from something that was external and operated mainly through coercion or absolute power (as in physical punishment) to something that is internalized, operated on the self and worked most powerfully on an ideological level through social relationships.

Foucault also referred to rituals and techniques in hospitals which established a power of normalization over individuals. Normalization is seen in 'the case' and the way in which it is 'described, judged, measured, compared with others' (1977: 191). In this way medicine socially constructs reality through its power to define what constitutes normality and therefore abnormality. Surveillance also extends to the body through the medical 'gaze' during the clinical encounter. This form of surveillance inscribes the body to such an extent that the individual starts to police or self-monitor her/his own body (Foucault 1976, 1980).

Foucault also described how the design of the hospital, like that of prisons and other large institutions, framed and helped to enact certain forms of bodily discipline and observation. The Panopticon, designed by Bentham for prisons, was circular with a central tower and prisoners' cells were positioned radially around the edge of the building. The central tower had a window and lighting system that allowed a supervisor to watch every prisoner in the building. Foucault asserted that this architectural design was utilized to insert this form of surveillance into other 'formidable disciplinary regimes' (1980: 58), for example factories, schools, hospitals, workhouses and army barracks. These institutions with their large populations developed specific hierarchies, spatial arrangements and surveillance systems centred on the requirement to supervise activities.

Following Foucault, a number of writers have described the industrial mode of production as a basic model for the modern hospital. Martin (1987), for example, framed her analysis within Marxist notions of the person's alienation and separation from the product of their labour. Martin (1987) highlighted the hegemonic connections between the hospitalization of women when in labour and birth and industrialization. Based on interviews with women about their experiences of labour and birth she presented a metaphor of the female body arguing that with the development of the hospitalized, complex management of birth, the woman came to be seen as part of an industrial factory. She argued that, in accordance with this model, the labouring woman is likewise disconnected from her birth, seeing it as something that is managed and controlled by the system. Women thus represented themselves as 'fragmented - lacking a sense of autonomy in the world and feeling carried along by forces beyond their control' (Martin 1987: 194). Thus, she asserted, women come to see their bodies as defined by implicit scientific metaphors.

Foucault's analysis of the design of hospitals and other institutions has been reflected in more recent developments in obstetrics, such as the use of electronic foetal monitors to enable midwives to observe a number of labours at a distance, from a central station, rather than staying with the woman in the birth room. Chapter 3, which looks at the active management of labour, examines the use of monitoring technology in marking and limiting the time of labour. Foucault's argument that prisoners or patients in such conditions internalized being surveyed and so began to internalize restraints or disciplines, which take the form of authoritative or hegemonic knowledge, are also relevant to the acceptance of monitoring technologies. As Arney (1982) argued, such technologies can be seen as limiting the autonomy of medical practitioners, as much as they can be seen as disciplining patients.

The sociologist Turner (1987) equally saw the development of scientific medicine in the nineteenth century as part of the medicalization of society, an aspect of the rationalisation of society through the dominance of scientific categories. As the work of Foucault and others has shown, and echoing Thompson's arguments about labour (Thompson 1967), people in modern Europe gradually became accustomed to and internalized different perspectives on the time and place of labour. By the late twentieth century, only a minority of women strongly questioned the idea that birth should not take place in its own time and in a domestic space.

This shift was led by North American and European societies, but was also spread worldwide through various globalization processes. One such case was the post-second-world-war Reconstruction in Japan, which saw the restructuring of health services along U.S. lines, based on North American assumptions about healthy lifestyles and behaviour and the value of medical care. Midwifery in Japan had been regulated in 1899 with midwives, as in the U.K., being independent practitioners caring for most births in the home or in small clinics (Hashimoto 2006). However, following the Reconstruction, their role was reconstituted along U.S. nurse-midwifery lines, with most births taking place in hospitals run along regimented lines.

Between 1902 and about 1930 in the U.K., bona fide informal midwives continued to work alongside the new generation of formally trained midwives, most still working in local communities and in women's homes, around the birth patterns of local women. They were still, in most cases, 'the woman you called'. From about 1936 in the U.K., with moves to make care more accessible to poor women, they became salaried workers, though still based in the community, and their work remained flexible to the rhythms of pregnancy and birth. In North America, midwifery was not formally regulated in the same way and, although midwives continued to practise in an informal way in rural communities – granny midwives, for example, in African American communities (Fraser 1995) – there was a rapid shift to hospital birth, managed by physicians and supported by nurses (Loudon 1992; Jordan 1993). In much of Europe, similar changes occurred but with domiciliary midwifery continuing for longer in some, such as the Netherlands (De Vries et al. 2001; Jordan 1993).

Following the Second World War in Britain, and with the inception of the National Health Service (NHS), a policy of providing hospital 'beds for all' was widely seen as a progressive aspect of modernity, aimed at producing greater social equality and likely to produce optimum health. However, the compromises with professional groups on which the NHS was founded left the hierarchical structures and disciplinary regimes developed earlier in the century in public hospitals firmly in place. Hospital wards were designed to maximize both professional observation and the perceived efficient movement of professional and caring functions in time.

Tew's (1990) critical history of childbirth in the U.K. highlighted how statistical evidence on 'risk' was, as with maternal death previously, misinterpreted and used to underpin a government policy of universal hospital birth, centralized into obstetric consultant-led units. As well as further declines in home birth, this meant the closure of many small and rural units run by midwives and general practitioners (GPs) with obstetric training. Along with this move in the place of birth came changes in practice. The use of developing medical technologies, which might have value for women with medical problems or birth complications, began to be normalized and treated as routine. Many interventions were performed in a routine fashion as aspects of the production line system in which labouring women needed to be 'moved along' the process of labour and 'delivered' efficiently. By the 1970s in much of the Euro-American world, interventions related to the time of labour – induction and augmentation techniques to start it and to speed it up – were used routinely and became the focus of protest by women's and midwives' groups. Despite the advent of consumerism in late modernity and successive reform policies, the active management of birth continued to be routine throughout most of the Western world. This is illustrated by accounts of the routine management of time in labour by British women in the 1990s (see Chapter 9, this volume).

During this period, the nature of midwives' work changed considerably. Once they were mainly working in hospital settings, midwives were essentially expected to staff the institution, rather than working around the pregnancy and birth patterns of women in their local community, in a flexible fashion with periods of relatively intense work or rest, not unlike the farmers described by Thompson (1967). Hospital midwives worked shift systems, organized in order to ensure the smooth running of a large hospital, with rotating shifts to ensure all ward areas were covered. In such a system they would care for many women for short intensive periods, during one particular aspect of their maternity experience (the effects of this on midwifery are discussed in Chapter 5, this volume). Using a Foucauldian-Marxist analysis this form of work can be categorized as essentially alienated with midwives performing a quasi-industrial worker role, obstetricians that of production management and women cast as a relatively passive part of the reproductive process (this is discussed further in Chapter 2, this volume).

Those midwives who continued to work in the community found their roles considerably narrowed and constrained, so that although not bound by hospital routines and shifts, they also cared for compartmentalized periods of pregnancy and postnatal care and rarely attended births. The system of shared care which was adopted in the U.K. to maintain the roles of GPs alongside hospital-based obstetricians, co-ordinated and shared out care in such a way that women's and midwives' experiences of childbearing were compartmentalized in a like fashion, with women moving around the system and between professionals in order to maintain its smooth flow across boundaries. Such disjunctures (from the woman's point of view) were touched on by Frankenberg (1992) in discussing the way time is shared or valued between professionals and patients in biomedical systems. He highlighted the paradoxical influence of time within biomedicine as 'patients' endure long periods of waiting interspersed by sudden intrusions upon their temporal and spatial boundaries.

Although we have focused here particularly on British history, such changes have been international. The historical processes through which these developments took place in northern Europe were spread through migration from Europe to America and through colonialism and global capitalism to less industrialized parts of the world. Chapter 7 discusses more recent moves in the north of Canada to restore aboriginal midwifery, which was progressively destroyed alongside European-origin midwifery with a policy of medically attended hospital birth for all women. Where this has occurred it has also been entrenched and exaggerated by colonial or postcolonial contexts of material and status inequality. The radical shift to medical childbirth which took place in North America was even more pronounced in Central and South America, fuelled by wide social and ethnic inequalities and rural–urban distinctions, so that by the 1980s Latin American countries such as Brazil and Chile had the highest rates of medicalized birth in the world, in urban settings and private hospitals, while poorer rural women continued to be attended by indigenous *parteras* and had limited access to medical care. Similar historical developments have been recounted in Middle Eastern (el-Nemer, Downe and Small 2006) and South Asian states (Donner 2004; Van Hollen 2003).

## Late Modernity and Postmodernity

As modernity arguably gives way to postmodernity, it might be argued that concepts of time in post-industrial societies are continuing to change in the face of late capitalism, globalization and consumerism. Theories of postmodernity appear to suggest that cultural conceptions of time may be returning in some ways to more fluid notions that were thought to characterize pre-industrial concepts. This is reflected, for example, in Stevens' account of caseload midwifery practice in Chapter 5 and Walsh's account of birth centres in Chapter 6. However, there are challenges for midwives in moving beyond the 'factory system' of time–work discipline and for women, such as the mothers in Chapters 9 to 11, who struggled to integrate elements of traditional as well as modern use and concepts of time, in a postmodern context. It could be argued that, in Western industrialized cultures, women have become so programmed by clock time that they may be unable to enter or experience cyclical time (Kahn 1989, Adam 1992). However, it is important to avoid dualistic representations of living in and with time, suggesting that we can only engage with one form of time or another. This emerged strongly in Hashimoto's accounts of Japanese motherhood in Chapter 11. Kahn (1989) also illustrated this with her experiences of motherhood and feeding which she argued allowed her to experience a more cyclical form of time in spite of living in clock/linear time. In particular she referred to her own experience of returning to work, where clock time predominated, and contrasts it with her experience of cyclical time when breastfeeding her baby:

> One of my favourite times of day was when I came home from work at 1 o'clock. I would lie down to nurse him off for a nap. After being at work, with deadlines, schedules and meetings, everything marked off by the clock, I would float with him into a different kind of time. It was more cyclical, like the seasons, the tides, like the milk which kept its own appointment with him without my planning it out. I lived during those years in two types of time - agricultural and industrial. I loved the two of them side by side. (Kahn 1989: 21)

This chapter has discussed historical changes which we suggest are particularly relevant to a situated understanding of childbirth and concepts of time. It has traced changes in how health care was provided, and how childbirth was brought in to the biomedical domain, alongside the wider developments in urbanization, industrialization and colonialism/capitalism that profoundly influenced, and were in turn influenced by, concepts of time. Historical analysis tells us that the concepts which we often take to be basic – given in the world – may be socially shaped, changing and changeable. When dealing with historical change in the present, there is a tendency both to 'naturalize' what we see as normal and to present changes in the form of oppositions, such as science versus nature, which then appear as though they are unchanging, within the process of change. Considering the processes of change in concepts and practices is useful in that it reminds us that the processes of healthcare, and changes in healthcare, are situated within a social context, and particularly that what is done is neither the result of an inevitable logic of scientific progress or rationality nor a product of natural laws. This leads us to consider the issue of cultural differences in concepts and management of time. The next chapter follows with a discussion of anthropological theories and concepts that may contribute to such an understanding.

## Notes

1. Laurence Sterne. 1978[1759]. *The Life and Opinions of Tristram Shandy, Gentleman.* London: Penguin, pp. 8–9.
2. *Guardian*, 22 November 2003.
3. The history presented in this chapter is necessarily a broad sketch, and the reader should refer to the many good histories of the relevant period, some of which are cited here, for more depth and detail. In its attempt to convey a general picture, such a sketch may appear to smooth out the great variety of experiences and the conflicting and contested nature of historical processes, but the complexity of change is acknowledged, and will be reflected further in the chapters to follow.

4.   None of the histories here suggest that the time period of the second may
      have bodily roots, in the beat of the heart, but such analogies have been
      drawn with music and the development of musical notation.

# References

Adam, B. 1992. 'Time and Health Implicated: A Conceptual Critique', in R.
      Frankenberg (ed.) *Time, Health and Medicine*. London: Sage, pp. 153–64.
Arney, W.R. 1982. *Power and the Profession of Obstetrics*. Chicago: University of Chicago
      Press.
Carter, J. and T. Duriez. 1986. *With Child: Birth through the Ages*. Edinburgh: Mainstream
      Publishing.
Cipolla, C.M. 1967. *Clocks and Culture 1300–1700*. London: Collins.
De Vries, R., et al. 2001. *Birth by Design: Pregnancy, Maternity Care and Midwifery in North
      America and Europe*. London: Routledge.
Dickens, C. 1999[1843]. *Martin Chuzzlewit*. London: Penguin.
Donner, H. 2004. 'Labour, Privatisation and Class: Middle-class Women's Experience
      of Changing Hospital Births in Calcutta', in M. Unnithan-Kumar (ed.)
      *Reproductive Agency, Medicine and the State: Cultural Transformations in
      Childbearing*. Oxford: Berghahn, pp. 113–36.
Donnison, J. 1988. 'The Decline of the Midwife', in *Midwives and Medical Men,
      A History of the Struggle for the Control of Childbirth*. London: Historical
      Publications Ltd.
el-Nemer, A., S. Downe and N. Small. 2006. 'She Would Help Me From the
      Heart: An Ethnography of Egyptian Women in Labour', *Social Science and
      Medicine* 62(1):81–92.
Foucault, M. 1976. *The Birth of the Clinic: An Archaeology of Medical Perception*.
      London: Tavistock.
——— 1977. *Discipline and Punish The Birth of the Prison*. Harmondsworth:
      Penguin.
——— 1980. *Power/Knowledge: Selected Interviews and Other Writings, 1972–1977*.
      London: Harvester.
Frankenberg, R. 1992. 'Your Time or Mine: Temporal Contradictions of Biomedical
      Practice', in R. Frankenberg (ed.) *Time, Health and Medicine*. London: Sage,
      pp. 1–29.
Fraser, G. 1995. 'Modern Bodies, Modern Minds: Midwifery and Reproductive
      Change in an African American Community', in F. Ginsburg and R. Rapp
      (eds), *Conceiving the New World Order: The Global Politics of Reproduction*.
      Berkeley: University of California Press, pp. 42–58.
Hale, J. 1994. *The Civilisation of the Renaissance in Europe*. London: Fontana.
Hampson, N. 1968. *The Enlightenment*. Harmondsworth: Penguin.
Hashimoto, N. 2006. 'Women's Experience of Breastfeeding in the Current
      Japanese Social Context: Learning From Women'. Ph.D. dissertation.
      London: Thames Valley University.
Heagerty, B. 1997. 'Willing Handmaidens of Science? The Struggle Over the
      New Midwife in Early Twentieth Century England', in M. Kirkham
      and E. Perkins (eds), *Reflections on Midwifery*. London: Baillére-Tindall,
      pp. 70–95.
Hobsbawm, E.J. 1968. *Industry and Empire: An Economic History of Britain Since 1750*.
      London: Weidenfeld and Nicholson.

Jordan, B. 1993. *Birth in Four Cultures: A Crosscultural Investigation of Childbirth in Yucatan, Holland, Sweden and the United States.* Prospect Heights, IL: Waveland Press.

Kahn, R.P. 1989. 'Women and Time in Childbirth and During Lactation', in F.J. Forman and C. Sowton (eds), *Taking Our Time: Feminist Perspectives on Temporality.* Oxford: Pergamon.

Kilaru, A., et al. 2004. 'She Has a Tender Body: Postpartum Morbidity and Care During Bananthana in Rural South India', in M. Unnithan-Kumar (ed.) *Reproductive Agency, Medicine and the State: Cultural Transformations in Childbearing.* Oxford: Berghahn, pp. 161–80.

Leap, N. and B. Hunter. 1993. 'Handywomen: the Women You Called For', in The Midwife's Tale: An Oral History from Handy Women to Professional Midwife. London: Scarlett Press.

Le Goff, J. 1986. *Time, Work and Culture in the Middle Ages.* Chicago: University of Chicago Press.

Loudon, I. 1992. *Death in Childbirth: An International Study of Maternal Care and Maternal Mortality.* Oxford: Clarendon Press.

Martin, E. 1987. *The Woman in the Body: A Cultural Analysis of Reproduction.* Milton Keynes: Open University Press.

Marx, K. 1970. *A Contribution to the Critique of Political Economy.* Moscow: Progress Publishers.

Mathias, P. 1983. *The First Industrial Nation: An Economic History of Britain 1700–1914,* 2nd edn. London: Methuen.

McGuire, S. 1998. *Visions of Modernity. Representation, Memory, Time and Space in the Age of the Cinema.* London: Sage.

Newell, R. 2004. 'The Thanksgiving of Women After Childbirth: A Blessing in Disguise?' Ph.D. dissertation. Dundee: University of Dundee.

Pawson, E. 1979. *The Early Industrial Revolution: Britain in the Eighteenth Century.* London: Batsford.

Stengers, I. 1997. 'Power and Invention: Situating Science', Minneapolis: University of Minnesota Press.

Sterne, L. 1978[1759]. *The Life and Opinions of Tristram Shandy, Gentleman.* London: Penguin.

Tew, M. 1990. *Safer Childbirth? A Critical History of Maternity Care.* London: Chapman and Hall.

Thompson, A. 1973. *The Dynamics of the Industrial Revolution.* London: Edward Arnold.

Thompson, E.P. 1967. 'Time, Work-discipline and Industrial Capitalism', *Past and Present* 38: 56–97.

——— 1968. *The Making of the English Working Class.* London: Penguin.

Towler, J. and H. Bramall. 1986. *Midwives in History and Society.* Beckenham: Croom Helm.

Turner, B. 1987. *Medical Power and Social Knowledge.* London: Sage.

Van Hollen, C. 2003. *Birth on the Threshold: Childbirth and Modernity in South India.* Berkeley: University of California Press.

Versluysen, M.C. 1981. 'Midwives, Medical Men and "Poor Women Labouring of Child": Lying-in Hospitals in Eighteenth Century London', in H. Roberts (ed.) *Women, Health and Reproduction.* London: Routledge and Kegan Paul, pp. 18–49.

Vincent-Priya, J. 1992. *Birth Traditions and Modern Pregnancy Care.* Shaftesbury: Element Books.

Weber, M. 1930. *The Protestant Ethic and the Spirit of Capitalism.* London: Allen and Unwin.

Wilson, A. 1995. *The Making of Man-midwifery: Childbirth in England 1660–1770.* Cambridge, MA: Harvard University Press.

Young, J. 2007, *The Vertigo of Late Modernity.* London: Sage.

## CHAPTER 2

# COSMOLOGIES, CONCEPTS AND THEORIES: TIME AND CHILDBIRTH IN CROSS-CULTURAL PERSPECTIVE

## Christine McCourt

Groups in all societies have devised interventionist rules and regulation to guide and control the behaviour of women in childbirth. These are matters which are almost never left to the discretion of the birthing women or to nature. And though these cultural prescriptions may differ between societies, all are guided by some rules and regulations. The crucial point is, of course, that – whether traditional or modern – all childbirth behaviour is culturally patterned.

—Lowis and McCaffrey (2005: 11)

## Introduction

The sociologist Adam (2004) has set out how religious philosophers through the ages have grappled with concepts of time. Time is present in origin myths – about the birth of humans in the world. She observed how successive world-views and modes of thought have understood time variously as universal and unchanging – a constant – and as ever changing, from past through present to future.

The anthropologist Bloch (1989), meanwhile, has focused on basic anthropological questions about the universality or relativity of human thought and experience, asking whether human cultures are, at a basic level, universally the same while the detailed beliefs and practices of cultures are extremely varied. This argument can aptly be applied to childbirth, which is universally physiological but also universally a cultural matter. All cultures are deeply concerned with their physical and social reproduction, but the concerns may be expressed in differing concepts and practices. Looking at childbirth concepts and practices cross-culturally anthropologists and others have found many echoing themes, as well as variations.

Anthropologists have always been interested in the margins of life, for what it can reveal about human beliefs – cosmology and ontology – and practices. However, this interest was focused traditionally far more on death than on birth. Possible reasons for this were discussed in the introduction to this book. However, the number of ethnographic studies concerned with childbirth and reproduction has grown in recent years. Additionally, a number of anthropologists and writers on childbirth have looked at the meaningful connections between life and death, since each are at the boundaries of physical and social life and are marked by rituals reflecting their fundamental importance. These boundaries also blur sometimes in situations where the beginnings and end of life merge, in a rupture, such as miscarriage, stillbirth or infant death, where the untimeliness of birth and death can lead to ambiguity about social personhood. It is also found in work focused on the life course and life transitions. While Judaeo-Christian cultures have tended to regard the life course as linear, life having a clear beginning and end, cultures influenced by other cosmologies have often viewed the life cycle as one where birth and death, literally or metaphorically, meet.

This chapter discusses the ways in which anthropologists have looked at childbirth, and at time, but also focuses on the theoretical perspectives within anthropology which can contribute to a broader understanding of beliefs and practices relating to childbirth. It looks first at anthropological studies of time, and then the anthropology of childbirth, and ways in which these connect in several ethnographies. It is intended to provide an introduction to thinking anthropologically about childbirth, and so to ask new questions, by making what may be familiar, to some readers at least, strange.

## Anthropological Theories and Perspectives on Time

We will start by briefly examining some of the major thinkers who have influenced anthropological theory, and who have also demonstrated a strong theme of time in their work, namely Marx, Weber and Durkheim.

Marx's (1906) theory of the commodification of labour, which relates time to the value of labour, was discussed in the previous chapter. Although his theories did not explicitly focus on time, it played a central conceptual role in his work through his focus on historical change and development, and particularly his analysis of capitalist development, labour and time compression. In Marx's analysis, the conversion of time to money through labour meant that, with the development of capitalist production, time – the way it is experienced and managed – became more and more compressed by the pressure to control work processes, to accelerate and maximize production. Although this work drew strongly on that of previous philosophers, it had a far more materialist and practice-oriented focus. Unlike those before him, such as the philosopher Kant, who saw time as having an a priori nature, Marx saw time as both formed by, and formative of, historical processes. His ideas were influenced by Hegel's development of the notion of dialectics as a resolution of dualisms in thinking about time – that is, that apparently opposite or dichotomous categories such as the social and the natural are in a continual interplay. Adam described Hegel's theory of time as 'immanent, as internal to and constitutive of living and social systems' (Adam 2004: 35) and so historical, but also as external (and so external to these and beyond history). Marx's theory of dialectics was one of continual interaction between thought (or ideology) and the material world, each influencing the other in a historical process. This theory is reflected in large areas of anthropological work, which focus on the relationship and interplay between cultural and material (or environmental) influences on human lives. Marx's theory of labour, in which the control, regulation and exploitation of time are central to capitalist development, was reflected particularly in the historian E.P. Thompson's (1967) study (discussed in Chapter 1, this volume). It is echoed again in recurrent themes in this book – such as the time compression experienced by aboriginal Canadians following settlement and involvement in wage labour (see Chapter 7). Adam noted how Marx argued that for time to be exchanged (converted, transacted) into the abstract, quantitative concept money, it too had to be abstracted in order to be valued and measured in quantitative terms, 'not the variable time of seasons, ageing, growth and decay, joy and pain, but the invariable, abstract time of the clock where one hour is the same irrespective of context or emotion' (Adam 2004: 38).

Adam argued that Marx's theory suggested that time is embedded in the various technologies and economic relations of capitalist society and 'as such, subject to conflicts that arise at the interstices of the different spheres: nature, society, home… each may have a different, not entirely compatible logic' (2004: 40). This observation raises questions about who has the power to impose a particular temporal structure as the norm. The relevance of these general theoretical observations about time, cultural norms and power to epistemologies of childbirth will become apparent through the chapters of this book, as they highlight and explore authoritative knowledge, discourses and practices in medicine and how these set norms of time and space in childbirth.

Weber's theoretical work was also deeply concerned with time and with power, although his focus was rather on the structures and bureaucratic forms of power than its relation to material conditions. His work on the Protestant ethic and the spirit of capitalism (Weber 1930) centred on the argument, like those of Marx, that ways of thinking about and managing time are socially and historically shaped. He noted that the dominant Christian notions of time – such as life having a clear beginning and end, albeit with possibilities of resurrection following the day of judgement – were different from those found in other religious systems. He traced the roots of this shift away from Old Testament, Judaic views to monastic lifestyle and asceticism, with the desire to overcome impulses from the world of nature, which included strict regulation of the body and attention to the hours of rising or sleeping, devotion and prayer. This was later transformed, with the Protestant Reformation to a more worldly form of asceticism, with an individualized form of self-discipline, which became very visible in the Protestant virtues of punctuality and the productive use of time. Gradually, as European society became more secular, the Protestant focus on calculability and control shifted from time-keeping to time accounting, trading and rationing. Weber also argued that the shift from Catholicism to Protestantism allowed a different orientation of past, present and future – towards a more future oriented, unidirectional and irreversible sense of change through time – once the idea of the redemption of past sins was lost.

Durkheim's equally influential theories emphasized the central role of the social in shaping human beliefs and actions. His notion of 'collective representations' (Durkheim 1965) represented a radical counter argument to the rationalist and utilitarian philosophies which had dominated much of nineteenth century European thought. His theory was that social experience and norms profoundly shaped cognition, to such a degree that concepts of time are fundamentally social. In Durkheim's socially constructivist approach, collective representations are both caused by and causative of social forms. Gell (1992) has cautioned that this approach represented a strongly relativist position, so that Durkheim effectively had to sidestep the issue of universal experiences of time, such as those grounded in bodily and astronomical experience. However, he also conceded that, taking a socially constructivist perspective, Durkheim saw even such basic experiences as those of the body as fundamentally socially shaped: society provides the organizing principles through which nature and sensory information can be understood. Therefore, while he did not deny that the external universe objectively exists, he argued that human experience of it is subjective.

Gell observed that in this respect, Durkheim's thinking (Durkheim 1951, 1965) drew particularly on the philosophy of Kant, which distinguished phenomena – representations, based on the world of sensory experiences – and noumena, underlying definitive laws of the world. Durkheim did not deny that time and space existed on the noumenal level, but saw it as being understood by humans on the phenomenal level. Gell (1992) argued convincingly that although Durkheim's emphasis on the social was over-stated his sociological arguments had value in highlighting the role of society in shaping thought, and in opening up new lines of enquiry about concepts of time.

Durkheim's ideas had a strong influence on anthropological theories of time in the twentieth century, particularly amongst structural-functionalist and structuralist thinkers. A major contribution to ideas about concepts of time, as well as concepts of health and illness, came from Evans-Pritchard's work in Africa. Following Kant and Durkheim, Evans-Pritchard (1940) distinguished 'oecological' notions of time – those derived from the environment and human interaction with it – and structural time – those geared to the organizational forms of the social structure. Gell (1992) argued, however, that in Evans-Pritchard's thinking, neither form is purely environmental or social, since it is in social groups and forms that humans respond to their environment, but these structures are in themselves influenced by the material forms of existence. This approach, although not explicitly so, echoed Marx's theory of dialectical materialism as well as Durkheim's social constructivism. Each attempts to relate the micro-social world of everyday experience to its macro-social context. At the micro level time is 'concrete, immanent and process linked' while 'the social processes by which macro-cosmic time is calibrated are themselves abstract, rather than concrete' (Gell 1992: 17).

The structuralist anthropologist Levi-Strauss, who was influenced by Durkheim, attempted to distinguish what he termed 'hot' and 'cold' societies in terms of their time orientation and concepts (Levi-Strauss 1969). Cold societies (by which he usually meant small-scale, less technologically developed societies) were characterized by social systems which produced an illusion of timelessness. Levi-Strauss saw social rituals as a means by which societies were able to overcome the contradictions between the 'synchrony' of such models and diachrony, the reality of change and the passing of time. Historical rites, for example, bring the past into the present, death rituals the present into the past, and 'rites of control' adjust the 'here and now' situation to the fixed scheme of relationships established in the mythic past (Gell 1992).

Critics such as Gell have argued that Levi-Strauss, like other anthropologists before him, tended to overplay the differences between societies in this respect, perhaps being too influenced by those societies' own ideological illusions of timelessness. Gell discussed the *Ida* ritual of the Umeda (a small-scale economically subsistent New Guinea society) in which time is ritually reversed, enacting 'a process of biosocial regeneration' (1992: 46) to 'manipulate processes in a symbolic way in order to indicate a certain normative path for events, thereby reinforcing the Umedas' confidence in the viability of their society' (1992: 53). The *Ida* ritual occurred annually, 'forming the temporal focus of the Umeda year' (1992: 38), and formed an important symbolic aspect of production (fertility of sago) and reproduction (fertility of people), which were viewed by the Umeda in a cyclical way. The dancers in the rite appear in age categories, from eldest first to youngest, their costumes and movements effectively acting out the life cycle in reverse order. He argued this did not mean that the experience of time per se is reversed, rather that the idea of cyclical time attempts to resolve the metaphysical conflict between time as a continuous process (diachrony, the passing of time) and synchronic oppositions (classifications such as old or young) and between process and structure. The ritual acting out of the life cycle attempts to ensure that diachrony 'only throws up the normatively approved kinds of events, as defined by the classificatory scheme' (1992: 53). This argument reflects a view of the function of ritual as symbolically enacting and conveying social norms, which is widely found in anthropological theory, and has also been applied to analyses of ritual processes in both biomedicine and other cultural forms of medical or healing practice. A good example of this is Davis-Floyd's analysis of U.S. hospital birth as a ritual which, she argued, enacts and confirms core values of American society in women (Davis-Floyd 1994a, 1994b).

Anthropological research, therefore, has important implications for concepts of time and how far these are regarded as relative or universal. In this book, we suggest that culture has a profound role in shaping ideas and practices, including those of medicine, midwives and women giving birth, but we also see an underlying logic to the patterns that all cultures share, just as they share the 'objective world' which Kant described as noumenal. Bloch (1989) has criticized the more extreme forms of cultural relativism in anthropological theory for confusing ideology with cognition, as well as criticizing some anthropologists for emphasising the ritual and symbolic aspects of culture over the mundane. He argued that cultures which are hierarchical give particular emphasis to ritual approaches to time, with ritual forms that tend to obscure the realities of power. In such contexts, ideological forms of time, such as a ritual calendar, come to dominate practical and embodied forms of time. James and Mills (2005) noted that much earlier anthropological work created an image of social time concepts as either 'modern linear' or 'traditional cyclical' but these are ideal types, not found in their totality in any society. In later chapters in this volume, we will see how examples of models of both linear and cyclical time are found within 'modern' societies, and therefore in their healthcare systems, and examples of how ideological forms of time came to dominate practical and embodied forms.

Nonetheless, social anthropology has continued to note the tendency of some cultures to privilege, or emphasize cyclical models of time, while others hold more linear progressive models. While some theorists, such as Bloch, have argued that cyclical models of time represent a ritual form, are characteristic of societies with hierarchical ideologies, and are designed to maintain them as such, others have pointed out that interaction with nature is itself a basis for cyclical views of time, because of seasonality and the demands of agriculture, and that ritual cycles are also likely to be related to these. Gell (1992), for example, compared views of time amongst the Umeda, a society reliant on hunter-gathering, and the Muria, an agriculturally based society. The Umeda had rather diffuse views and terms for time. They didn't have words for months and days, although they did have means of reckoning and counting them, and their ritual cycle was focused on one major annual ritual. In contrast, the Muria had an elaborate ritual cycle and strong focus on calendars, which perhaps 'epitomise the essential – temporal – form of the farmer's predicament, offering a magical surrogate for control over time and chance which the peasant, always on the horns of some planning dilemma, never has' (1992: 89). Gell argued that faced with such dilemmas, 'different societies or social strata, operating under different ecological circumstances, employing different technologies and faced with different kinds of long-term and short-term planning problems, construct quite different cultural vocabularies for handling temporal relationships' (1992: 89).

Similarly, Dilley (2005), writing of the Haalpulaaren of Senegal, argued that differing forms of cyclical time and non-linear representations of the past and present were variously crucial to magical practitioners who do not have high social status, to the Muslim calendar, to agricultural cycles and to genealogical reckoning, all of which formed aspects of local concepts of time. In the case of the Muslim calendar, he argued that it was based on both cyclical (repetitive) and on teleological (chronological) historical notions of time, which appeared designed to bring about the growth of Islam.

Gell (1992), in his extended essay on time, attempted to construct an anthropological theory of time cognition that overcomes the criticisms of both strong relativism and strong universalism. Drawing on the ideas of the philosophers Mellor and McTaggart, he focused on the two philosophical categories of time, which philosophers have named A-series and B-series time, which have been debated considerably within philosophy. A-series time represents the notions of tense – a sense of past, present and future – while B-series time represents the notion of order – a sense of before and after. Gell argued that much can be explained in terms of these two fundamental categories of time experience. He proposed that the A-series and B-series can be compared with the categories outlined by earlier thinkers (such as Kant's phenomena and noumena) and also with the work of phenomenologically inclined thinkers such as the social psychologist G.H. Mead and the philosopher Husserl. The A-series is inevitably dynamic, since any event or experience must pass from a future state through the present to the past. Humans continually and actively revise their notions of past, present and future as time passes, whereas the concepts of order appear fixed, enduring. Phenomenology focuses on the A-series notion, which fits well with its emphasis on lived experience and its more relativistic stance.

However, according to Gell, B-series time is basic because it doesn't change: any event has a date and this is universal and unchanging, even though the indices used to mark, measure or describe dates may vary culturally. He suggested that all humans operate with both models of time. They must have cognitive maps of time, which can be distinguished from our (phenomenal) day-to-day beliefs about what the situation is now. He looked at the work of Husserl, on the psychology of internal time consciousness, and the cognitive psychologist Neisser, who built on Husserl's ideas, to attempt to build a bridge between relativistic and universalistic theories. Husserl developed the idea that present perceptions are shifted into the past through a process of continual modification via 'retentions' (from the immediate past) and 'protentions' (anticipations of the future). Neisser argued that perception depends on matching an input with a stored schema, and that the person's stock of stored schema (from past experience) are also modified in the light of experience. This suggests a cyclical, constructive model of cognition. Gell argued, using the idea of schema, that 'perceptual images are mapped onto the corpus of B-series belief inscriptions, which form the basis of mental maps of time' (1992: 235). In his view, time is both universal and culturally varied, an argument resting on a 'distinction between time and the processes which happen in time'; 'there is no fairyland where people experience time in a completely different way. There are only other clocks, other schedules' (1992: 315).

The anthropologist Bourdieu (1977) also attempted to break down and bridge the extreme theoretical oppositions of subjectivism and objectivism. His concept of 'habitus' focused on practices and their relationship with structures, arguing that knowledge and patterns of behaviour are developed in interaction with the environment as well as by reference to culturally acquired schema (or dispositions). These dispositions are themselves generated in the objective conditions of their environment. Dispositions are deeply embodied because they are 'schemes of perception and appreciation deposited, in their incorporated state, in every member of the group' (Bourdieu 1977: 18).[1]

Bourdieu's theory of habitus, Gell argued, keeps what is useful about structural theories – the capacity to focus on systems as systems – without falling into the trap of seeing cultures as fixed entities, unchanging, and individuals as simply passive actors within a particular cultural scheme. In line with Marx's concept of dialectics, the concept of habitus rests on the view that social actors actively form their ideas and dispositions in interaction with their environment, but they do this in accordance with cultural influences which are laid down in their earliest upbringing, and which have even deeply shaped the environment they experience and act upon. Since they are often deeply embodied, these are not necessarily explicit or consciously followed, like rules, but tend to form internalized structures and schemes of perception which nonetheless guide what people think and do.

James and Mills (2005), in an edited collection of anthropological essays on time, noted the validity of critiques of earlier anthropologists for creating an illusion of timeless societies. However, they argued that the work of early anthropologists –such as Van Gennep, on life cycle rites of transition, and Hertz, on the social mapping of the body – have had enduring resonance and value. Van Gennep's (1960) theory of rites of passage involved three distinct phases – of separation, liminality and reincorporation in a changed social state – and has been widely applied, helping make sense of how changes in the life course are embodied and socially incorporated. At key stages of life, or social transitions such as reaching adulthood or becoming a parent, in a range of cultures, rites of passage separate the person from their normal lives and environment, placing them in an isolated ritual space reflecting their 'betwixt and between' status, where they often undergo ritualized learning or tests, before being literally and symbolically returned to the ordinary social and physical world with a new social status and role. Van Gennep's work was used by Davis-Floyd (1994a & b), for example, as a model for her analysis of U.S. hospital birth as a ritual, except in her argument the rituals ignored the significance of birth in the woman's personal life cycle and were focused on the needs of the institution, and the woman's transition within the expectations of U.S. culture. Similarly, Hertz (1960) discussed how the social order is mapped, metaphorically, onto the body, and Douglas's (1973) influential writing on the body and symbolism described how in many societies the individual body is regarded as a microcosm of the social body. For example, in analysing the Old Testament book of Leviticus, she analysed the Judaic ritual analogy between the structure of the body and the structure of the temple, the temple itself being an analogical model of the cosmos (Douglas 1999).

While agreeing with his dissatisfaction with strong versions of relativism, James and Mills criticised Gell's own theory of time cognition as being too phenomenologically focused on the 'here and now' and on the importance of 'opportunity cost' reckoning, which has been particularly associated with the modern 'protestant ethic' of time: for example, that people universally weigh up time like a resource that if spent in one way cannot be spent in another. They discussed the qualities imbued in time and calendars, arguing that the object of a calendar is not to measure time (as it might usually appear to us) but to imbue it with rhythm. Time is thereby endowed with a social nature. They suggested, therefore, that it is important for anthropological studies to attempt to find connections between (socially) conventional representations of time and the ways things are actually done. 'The real challenge', they argue, 'is to distance ourselves intellectually from our current, and perhaps local, sense of the shapes of time, including our current ideas of "modernity" and to place these alongside other senses of the shapes of time which have informed human life, action and experience' (James and Mills 2005: 6).

In the same volume, Hsu (2005) looked at practices associated with Chinese traditional medicine, alongside African traditional and Western biomedicine, to challenge the temporal assumption (found in Gell's characterization of B-series time – order – as objective and unchanging) that diagnosis always precedes treatment. In practice, she found that in all cases the time order is not so easily defined, and processes of diagnosis and treatment are often intertwined, with understanding emerging from interaction with the body of the patient during treatment. She cited comparable examples from Young's work on Post Traumatic Stress Disorder (Young 1998), and the well-known tendency of particular disease diagnoses to increase once treatments for them become available. She argued, therefore, that far from being objective and culture free, as Gell assumed, unidirectional time vectors are used by ethnographers as an ordering device.

Heintz (2005) discussed changing notions of time in the process of social change from socialist to capitalist Romania, arguing 'time was an efficient political tool and the "colonisation" of individual time was an important way of counteracting possible acts of resistance' (2005: 175). In socialist Romania, with an 'economy of shortage', continually being made to wait dominated people's experiences of time and its predictability. With the shift to capitalism the concept of time as money came to take hold more, but she found that in the workplace people had difficulty in translating this changing concept into practice. Furthermore, she observed that most Romanians are able to, and do, move freely between urban life, where such concepts have become more established, and rural life with a more seasonal, task-focused approach characteristic of Thompson's (1967) account of pre-industrial time concepts (see Chapter 1, this volume). Clearly, she noted, people don't have two completely different, incommensurate, concepts of time, but the capitalist measure of time primarily in terms of hours, rather than tasks done, is imposed on workers who used to understand time in this way. She also suggested that the history of Romania, of being dominated by outside powers and circumstances, has encouraged a relatively fatalistic view, with consequent effects on work practices and values, as well as Romanians' sense of self.

Mills (2005), in an account of Ladakhi (Tibetan Buddhist) time concepts, attempted to look at time as a system of power in itself. Reflecting on the criticisms of Geertz's account of time in Bali as being too taken up with ritual, ceremonial representations of time (for example, Bloch 1989), Mills noted that the Hindu calendar has an elaborate cyclical ritual nature which refuses to see duration or historical time as salient. Embodied practices are used to integrate individuals into the general time scheme, rather than requiring a sort of abstract engagement. He argued that the unreflective, embodied integration of actions in relation to standardized clock time in European time is essentially similar. He argued that 'time anthropology' therefore involves two linked considerations:

> firstly, a more or less abstract examination of *why people represent time in a particular way* (and who might benefit from them doing so; and secondly, an examination of *the embodied practices by which people do time* – how they orient themselves towards particular temporal/calendrical ideologies and thereby integrate themselves into wider ideologically-structured communities. (Mills 2005: 350)

## Anthropological Studies of Childbirth

Rituals involve embodied practices and experiences, as well as drawing on cultural schemata: they are performed and learnt by the body rather than being simply abstract ideas. In this way, the practices are in a sense inscribed on the body, conveying meanings more deeply than more abstract forms (such as a creed or ideology) alone could. Mills (2005), discussing birth (and death) practices in Ladakh, argued that ritual action integrates individuals into a particular time scheme powerfully in this way.[2] As noted above, Ladakhi Buddhists have a series of quite elaborate ritual and household calendars, based on the lunar month, the solar cycle and astrological observations. Concepts of pollution are also an important aspect of ritual and practical activity. The reproductive body of the household is linked to sources of fertility and those who can control them through sacrificial modes of exchange. He argued that the temporal disruption of the ordinary course of events brought about by birth or death creates a source of pollution. In Ladakhi terms, the fetus – a person not yet born – is linked to the water spirits (*lu*) who influence fertility in general. The separation of the fetus from the mother's body during birth is seen as a radical severance. The pollution emanating from such disruption is seen as a barrier to social interaction, since it places the mother and father outside normal ritual time, producing 'a kind of "shadow" or "demonic" time' (Mills 2005: 360) that flows alongside ordinary social time, and is inauspicious because those concerned are cut off from ordinary personhood. For the first seven days, father, mother and baby are secluded, during which period they can only be visited by male members of the household. Following an initial set of purification rites, the father is allowed to leave the house and others can visit the house to bring gifts and food but not eat there. Following a further thirty-day rite, the house returns to normal functioning and guests can eat there and touch the baby. Mills argued that 'the performance of purificatory rites thus "re-inserts" individuals and households back into a nominal hierarchy from which they have been temporarily dislocated' (2005: 360).

This theme of disruption in the life course by events such as birth, and the need to reintegrate people into ordinary social time, reflects Van Gennep's original argument about the importance of life-cycle rituals. The person, when ritually separated, is out of ordinary social time and space and is often seen as especially vulnerable. In her discussion of concepts of purity and danger, Douglas (1966) commented on how in a number of societies the fetus is seen as both vulnerable and potentially dangerous in this way. The unborn baby is considered to be in a liminal or transitional state and does not yet fit into the regular temporal scheme of things.

Van Hollen (2003), in an anthropological study of childbirth among poor women in Tamil Nadu, focused on how women have responded to social changes such as increasing medicalization, a shift toward birth in hospitals and the international development agenda, which includes control of reproduction. They were very critical of hospital maternity care, particularly due to its discriminatory and unsupportive nature, but were keen to use certain selected aspects of the new medical technologies, particularly induction and augmentation of labour by an artificial hormone drip; in other words, managing its timing and speed.

An illustration of the change in social and cultural practice connected to pregnancy over time was Van Hollen's observation of the reinvention of the *Cimantam* ceremony, which marked the progress of the pregnancy by ritual at key months. This was generally held for women in their first pregnancy, and said to mark its auspiciousness as well as to satisfy the *acai* ('cravings' or 'desires') thought to emanate from the fetus and the mother's pregnant state (2003: 77). As the Tamil calendar is lunar, pregnancy lasts ten months, and the *Cimantam* would generally be held in the ninth, but sometimes also in the fifth and seventh month of pregnancy. The fifth month was reported to be the time at which life becomes associated with the fetus, generating its *acai*. Where the family could afford this, Van Hollen noted that the ceremony, which had often been very simple, could involve an elaborate demonstration of consumption, and informants reported this was increasingly the case (2003: 77–78). They argued that the *Cimantam* was a relatively new tradition, ushered in with independence from British rule, but she observed that it was clearly rooted in Brahmanical traditions and may have been performed more exclusively before this among higher caste families. She suggested, therefore, that the tradition has been reinvented, in a more 'modern' form, less restricted by caste but restricted (or expanded) by socio-economic status (2003: 97). The conspicuous consumption involved perhaps reflected the new economic India as well as older traditions. The case of *Cimantam*, therefore, shows how shifts in practices associated with pregnancy and childbirth over time are related to social and cultural changes, often in very complex ways. Similarly, it shows how the concept of 'time' and the practices surrounding it are embedded in complex social processes. In this context, pregnancy is marked out by time in particular ways – the number of months and significance of key points – which we will see differ from those taken for granted as natural and given periodizations in biomedicine.

In talking to women, Van Hollen found their attitudes to labour pain were very positive, indeed, that they saw strong 'grinding' pain as a normal and necessary part of labour (2003: 112). This unexpected view (given her own familiarity with many U.S. women's dread of labour pain) was elicited in response to her questions about whether women were given medication for pain: the Tamil women welcomed the possibility that the drips and injections might make the pain stronger. She found a strong preference for acceleration of labour, highlighting that medicalization may take place in different ways in different contexts and that women are active negotiators, rather than simply passive victims of change. The concept of female power – *sakti* – was seen as rooted in women's reproductive capacity and ability to endure the pain of birth, making women valiant and able to cope with birth, and giving birth also enhanced this power. However, some women said that *sakti* had decreased in modern times with changes in women's lifestyle, affecting their power in childbirth, and making interventions such as using (oxytocin) drips more necessary (2003: 122). Some also felt that the use of such injections, to increase the power of contractions and speed up labour would in itself enhance the weakened *sakti* (2003: 138). Similar views of modern women's declining power to labour and give birth have also been noted among African-American women (Fraser 1995), and among Mayan women in Yucatan (Jordan 1993). In later chapters in this book we describe how time in labour is actively managed in biomedicine, and the ways in which this may be resisted by midwives (Chapter 4) or may conflict with women's experiences of time (Chapter 9). This case, however, highlights that even in 'traditional' or 'non-Western' settings, women will not necessarily view the speeding up of labour as an oppressive intervention.

Van Hollen described a complex set of influences on this selective acceptance of biomedical intervention. First were the very real memories of deaths of women and babies through obstructed labour in times before women had any access to medical emergency care if needed. Herbal preparations, usually ginger-based, were traditionally used to enhance and bring on labour contractions, and were still in regular use, but seen as less potent than the allopathic drips or injections (2003: 125). Further notable influences were the overcrowded and poor conditions of labour wards, encouraging staff to process women through as quickly as possible, and the role of the pharmaceutical industry in promoting the use of oxytocin injections. However, the overcrowded and busy conditions, with rushed staff tending to either leave women alone in a way they feared, or to mistreat them if they cried in pain, also made women very reluctant to use the public hospitals available to them. Similar findings have also been reported by D'Ambruoso, Abbey and Hussein (2005) in Ghana, and for aboriginal women in Canada (see Chapter 7, this volume). Van Hollen commented that it remains to be seen what the long-term impact of these changes will be, whether women will increasingly come to be seen as lacking *sakti* and in need of medical interventions to bring about a timely birth and how this might, in turn, impact on other areas of life (2003: 140).

Looking to postnatal care, Van Hollen's ethnography focused particularly on the impact of international health agenda on hospital practices and women's experiences (2003: 166–205). For instance, in contrast to the U.S., where she observed women had to fight for rights to postnatal care, a stay of up to a week following birth was encouraged in Indian hospitals. However, in addition to the focus on rest and recovery, this flowed from the development agenda, which set strong targets to ensure that women accepted IUD insertion, to ensure proper time interval before subsequent births, or sterilization, and to ensure an appropriate number of births. Postnatal care, she argued, is a key site for the international development agenda and the manoeuvres of both the health professionals and the populations they serve around them. The period in hospital was also seen by workers as a time-limited opportunity to influence mothers before they returned to their home, where traditional influences would increase their resistance to change (2003: 176). As noted in Chapter 1, management of time in relation to childbirth, in this analysis, was deeply tied to concerns with social reproduction, with the body of 'the Indian woman' representing the health of the social body (2003: 49).

Thomas (1992) focused on three key points in women's lives where time plays a central role in ensuring an orderly transition: menstruation, conception and pregnancy/birth. Each have their proper social time and place, and involve passages across and the maintenance of boundaries.

She referred to Frankenberg's argument that the 'contradiction of temporalities and of aspects of body boundaries is sharpest between social and natural motherhood on the one side and class and gender-constructed obstetric medicine on the other' (Frankenberg 1992: 25). Thomas noted that the pattern of antenatal care itself was a temporal sequence whose intensity increases with advancing gestation. This clear pattern, of decreasing time intervals between visits, was set down in custom and practice in Britain, as a public health response to the poor condition of military recruits and high rates of infant mortality, but its precise form was never tested medically. She observed that such timing may not correspond to the woman's identification of her needs, and that since there are few antenatal interventions that can make a difference during pregnancy, most effort goes into surveillance. She argued that with estimation of the date of delivery, a primary function of antenatal surveillance, birth becomes not 'her time' (see Chapter 9, this volume) but 'their time' and is calculated using their methods. While women, on booking with the maternity services, are asked in many countries to give a medical history including the pattern, dates and duration of menstruation, this information is now collected but discounted in favour of ultrasound measurements which convert linear measurements of growth into time. As Thomas stated:

> What is measured is the foetus and therein lies the first of numerous translations of time into other calibrations. If, subsequently, labour does not begin spontaneously within two weeks of the calculated due date, induction will be used to ensure a timely delivery. During labour, the speed of dilation of the cervix is monitored, and if outside accepted norms, acceleration will be used. (Thomas 1992: 62–63)

She concluded, 'the cervix is the site of many of the transformation scenes of reproduction: in acting as gatekeeper, medicine finds its role to be that of timekeeper' (1992: 65).

Simonds (2002), in a discourse analysis of U.S. obstetric, midwifery and parent self-help texts from the 1970s to the 1990s, commented on how far 'obstetrics works on women's bodies to make them stay on time and on course' (2002: 559). The history of childbirth in the U.S. was one where obstetrics assumed a higher degree of control than in European countries, and midwives' roles became marginal and to some extent counter-cultural (Loudon 1992). High levels of obstetric control and intervention were advocated and accepted from about the 1920s onwards, and this was only influenced to a limited degree by the impact of feminist critique during the 1970s in America. Additionally, Simonds argued that while some reforms have had an impact in U.S. hospitals, time-based mechanisms of medical control have proliferated, often becoming less visible as they develop and become routine. She showed how texts on pregnancy focus on milestones and their precise measurement. The beginning of pregnancy, like the beginning of labour, should be precisely defined and measured by professionals. She noted how the milestones or time measures of pregnancy had shifted historically from trimesters or months to weeks, a more culturally derived (based on the notion of the working week) as well as more finely graded form of counting time.[3] Concerns with the duration of pregnancy also became more precise and compelling in the discourse over this period, with increasing emphasis on routine induction of labour in 'post-term' pregnancy. In labour, which is divided into three main stages, increasing incrementalization of time appears to provide a cognitive sense of order and control for professional attendants, but also produces a feeling of time as being scarce and in need of control (see also Chapter 9, this volume). This analysis could be viewed in terms of the Marxist theory of labour, applied to the more 'professional' role of health workers, as was discussed by Martin and Arney (see Chapter 1). The active management of labour based on Friedman's partogram (see Chapters 3 and 4) also hinged strongly on measurement and control of time. Simonds noted the logical confusion between statistical and physiological norms in the way such developments were applied to all women. Postnatally, she noted that texts go on to prescribe 'obsessively precise' timing and duration norms for breastfeeding, in the manner of prescriptions for administering medication. These themes are discussed in greater depth in subsequent chapters of this book.

In her work based on ethnographic research in an Italian hospital, Pizzini (1992) contrasted the 'interior time' of women's embodied and perhaps unconscious experience in pregnancy with that of 'social time-in-the-world'. From her research with women, she suggested that giving birth appeared as a rupture, not only in a physiological sense – one becoming two – or psychological sense – the transition to motherhood – but also as a product of the prevailing logic of hospital birth. When a woman experiencing regular labour contractions comes into hospital, there is a break in terms of place as well as time, and her pains often cease, only to restart later. If she does not go into labour at the expected time, pharmacological induction will be used. 'The woman may believe that this type of intervention is necessary for her own health and that of her child, but at the same time may recognise the violence being done and the doubtful utility of its application to each single case' (1992: 70). Pizzini built on the earlier work of Rosengren and Devault, who argued, from prolonged observation in a maternity hospital, that the structures of both time and space served primarily to delineate status and define roles. She looked at the rhythm of activities and noted that, in practice, things which were described by health workers as a necessity, immutable in time and space, in fact changed according to variations between the people involved in the interaction – for example, approaching the end of a shift, or between day and night shifts. Furthermore, interventions may be used in order to maintain norms of time – neither precipitate nor too slow – in the absence of a particular pathology. Although norms of time themselves may be used to infer pathology, as will be discussed in Chapter 9, these norms vary culturally and historically, and even between particular institutions and practitioners. Similarly, temporal, institutional organization takes place through the sequences of timing, so that the labouring woman's trajectory through them is scrupulously followed, almost regardless of the physiological trajectory of her labour.

From these observations, Pizzini argued that, evidently, the division between subjective and objective time reported by the women she interviewed could also be applied to the staff who attended them. She observed a general speeding up of time and activity around the delivery room, which was understandable given that birth is the culmination of a long and emotionally involving process, but was also added to by the presence of doctors, who are the main birth attendants in Italy.

## Anthropological Studies of Time in Medicine and Medical Power

Frankenberg (1992) argued that the creation and control of power through the medium of time is central to the cultural performance of sickness, and a central way that medicine creates its 'other' and makes its object: 'the rigid time structures of the hospital emphasise the anti-temporality of the experience in relation to "normal", worldly time' (1992: 23). Frankenberg also developed the theoretical approach set out by anthropologists such as Scheper-Hughes and Lock (1987) of the three bodies – the individual body, the social body and the body politic – as also found in Douglas's (1966, 1973) depiction of the 'natural' body and the 'social' body, reproducing and reinforcing reciprocal images of each other. Hence, the ways in which sickness is dealt with are revealing about social relations, as well as reflecting them. Responding to Kleinman and Eisenberg's (1981) writing on forms of sickness – disease, illness and sickness, corresponding to bio-medical, personal and social understandings – Frankenberg argued that the cultural performance of sickness always involves a use of time and space which is different from the everyday:

> the removal of the patient's body to the hospital implies more than its biological reshaping. In fact, the total cultural performance of biomedical healing is concerned neither with social persons nor with natural physiological or pathological processes but precisely with transitions between the two. The hospital body and the embodied patient are seen as bounded, and what has to be controlled are movements within and across these boundaries. In this control, time is the essential factor. (1992: 22)

The 'three bodies' or forms of sickness also reflect the analytical framework of the social sciences. These work at the micro-social level – the level of the individual and the local, at the level of social relationships and values – and the macro-social level of wider overarching social structures and processes. In medical sociology these theoretical levels of analysis are sometimes referred to respectively as the 'phenomenology of sickness', 'the sociology of the sick role' and the 'political economy of health' (Turner 1987: 220–21).

While Young suggested that sickness is 'a process for socialising disease and illness' (1982: 270), it might similarly be argued that Euro-American maternity care is a process that medicalizes the social as well as physiological aspects of childbirth. In maternity services, women are accepted as being healthy and well in most cases, yet there is nonetheless a tendency in biomedical systems to approach birth, if not pregnancy, as a 'sickness role'. When the sociologist Parsons (1951) set out the concept of the sick role he described it as a socially sanctioned role in which the patient is separated and removed from their normal responsibilities and rights, but is expected to fulfil specific expectations of appropriate illness behaviour. These include co-operation with medical professionals and adopting behaviours that are seen or prescribed as health promoting. As part of relinquishing usual social obligations, the sick person is expected to entrust some measure of power and control to health professionals.

To exercise such control, medicine is involved in the standardization of illness into categories that can be managed by bureaucratic agencies. Social control of sickness, of course, may not seem very relevant to pregnancy and childbirth, where a woman is usually healthy and undergoing a normal physiological and life-cycle process. However, as discussed, such ideas can be found in a range of studies that look at the ways in which women are treated in pregnancy and childbirth, and how these can be understood in terms of social control. The anthropologists Martin (1989) and Davis-Floyd (1994a, 1994b) have discussed how care for women in childbirth can work as a form of social control or a way of reinforcing core cultural values. Jordan (1993) has also discussed how promoting the institution of authoritative knowledge enables social control to work very effectively without the need for external coercion.

As discussed in Chapter 1, a number of writers, drawing on Marxist theory, have observed how the industrial mode of production – epitomized by Henry Ford's car production line – was used as a model for hospitals in the modern era. Hospitals were run using an approach that was characteristic of most factory production until the advent of reforms in the late twentieth century. Modern Western hospitals continue to use very similar models and these models have been spread globally, even in more resource-poor countries. Arney (1982) and Martin (1989), influenced by Foucault's critical analysis of power, have commented on how the roles and relationships between doctors, midwives and patients tended to fit with this model with the doctor perceived metaphorically as manager, midwife as worker, mother as machine and baby as product. While Martin particularly focused on mechanistic attitudes to the female body, Arney's 'ecological' analysis emphasizes the role of continual monitoring in the (self-) regulation of actions in time and space (see also Chapter 9, this volume).

Marx's concept of alienation of labour in production can also be applied in similar ways to reproduction. Martin (1989) described the dominant metaphor of the body in medical care as one of fragmentation, the separation of the self from the body and feelings of alienation. She

noted how the use of an industrial mode of production in U.S. labour
wards and its extension through the use of new technologies appeared
to remove the need for the labourers. It made pregnancy and birth even
more like commodity production. She noted the importance of social
relations of production in the uses of medical technologies. With recent
developments in medical technology, even the traditionally powerful
doctor is downgraded in this system from manager to worker. She
also described the prevalence in medical institutions of covert forms of
resistance (for example, among women and nurses) such as delaying
diagnosis of active labour in order to 'buy women time' to give birth
without intervention. This theme is also explored in depth in Kirkham's
(1989) analysis of relationships on the labour ward.

The anthropologist Arney (1982) developed a similar analysis of late
modern medical practice. He showed how commodity relations and the
organization of medical care encouraged the rise of monitoring technology
in medical practice, particularly in obstetrics, so that midwives and
doctors are reduced to labourers in the production process and women
are treated as passive in the process. Referring back to Foucault's work
on the dominance of continual observation in medical institutions, he
argued that all participants were under surveillance in the new medical
model of birth, rather than simply the patients, so that all are likely to
self-regulate or discipline their behaviour in the light of prevailing norms
and ideas.

Foucault (1980) saw power and knowledge as closely connected.
He highlighted the importance of enlightenment changes in the way
discipline operated. He argued that, in the shift from traditionalism to
modernity, discipline changed from something that was external and
operated mainly through coercion or absolute power (as in physical
punishment) to something that is internalized, operated on the self
and worked most powerfully on an ideological level through social
relationships. Panopticonism, as analysed by Foucault (1979), is inscribed
in the design of the 'Nightingale-style' ward with long rows of beds and
very little privacy which is still found in many 'modern' hospitals (see
Chapter 10, this volume). More recently, with technological developments
in obstetrics, it is facilitated through the use of electronic fetal monitors
which enable midwives to observe a number of labours at a distance, from
a central station, rather than staying with the woman in the (private) birth
room, something that requires time to allow for the lengthy attendance of
a woman in labour.

This theme is picked up in several of the chapters to follow, describing
childbirth in hospitals. An interesting aspect of Foucault's analysis was
that prisoners or patients in such conditions internalized being surveyed
and so began to enact self-restraints or disciplines. Arney's (1982) analysis
also highlighted how far this may apply to carers and professionals too.
This adds to the point made above and in Chapter 1 about the power
of hegemonic knowledge. Coercion becomes less and less important
in maintaining dominant ideas as these are internalized and operated

by people within the institution. In these chapters we will see how authoritative knowledge is both enacted and maintained, as well as resisted and changed, and how medical practices come to be normalized, by practitioners as well as users of health care, particularly through establishing ways of understanding and managing time and space.

## Notes

1.  The translated term 'incorporated state' should convey the idea of embodiment, since Bourdieu (1977) emphasized the embodied nature of learning and experience, both through cultural schema and practical action and interaction (practice).
2.  This argument is echoed neatly in Davis-Floyd's (1994a) account of the role of ritual in American childbirth.
3.  Simonds's argument, that this is based on the working week established in industrial societies, echoes Thompson's (1967) historical argument about the social impact of industrial capitalist development (see Chapter 1). The concept of a seven-day week arguably has older cultural origins in Judaeo-Christian societies, with the belief that God made the world in seven days.

## References

Adam, B. 2004. *Time*. Cambridge: Polity Press.

Arney, W. 1982. *Power and the Profession of Obstetrics*. Chicago: University of Chicago Press.

Bloch, M. 1989. 'The Past and the Present in the Present', in *Ritual, History and Power*. London: Athlone, pp. 1–18.

Bourdieu, P. 1977. *Outline of a Theory of Practice*. Cambridge: Cambridge University Press.

D'Ambruoso, L., M. Abbey and J. Hussein. 2005. 'Please Understand When I Cry Out in Pain: Women's Accounts of Maternity Services During Labour and Delivery in Ghana', *BMC (BioMed Central) Public Health 2005* 5: 140.

Davis-Floyd, R. 1994a. 'The Technocratic Body: American Childbirth as Cultural Expression'. *Social Science and Medicine* 38: 1125–40.

———— 1994b. 'The Ritual of Hospital Birth in America', in J. Spradley and D. McCurdey (eds), *Conformity and Conflict: Readings in Cultural Anthropology*. New York: Harper Collins, pp. 323–40.

Dilley, R. 2005. 'Time Shapes and Cultural Agency Among West African Craft Specialists', in W. James and D. Mills (eds), *The Qualities of Time: Anthropological Approaches*. Oxford: Berg, pp. 235–48.

Douglas, M. 1966. *Purity and Danger: An Analysis of Concepts of Pollution and Taboo*. London: Routledge and Kegan Paul.

———— 1973. *Natural Symbols*. Harmondsworth: Pelican.

———— 1999. *Leviticus as Literature*. Oxford: Oxford University Press.

Durkheim, E. 1965[1912]. *The Elementary Forms of the Religious Life*. New York: Free Press.

———— 1951[1897]. *Suicide. A Study in Sociology*. New York: Free Press.

Evans-Pritchard, E. 1940. *The Nuer*. Oxford: Clarendon Press.

Foucault, M. 1979. *Discipline and Punish: The Birth of the Prison*. Harmondsworth: Penguin.

——— 1980. *Power/Knowledge: Selected Interviews and Other Writings, 1972–1977*. London: Harvester Wheatsheaf.

Frankenberg, R. 1992. 'Your Time or Mine: Temporal Contradictions of Biomedical Practice', in R. Frankenberg (ed.) *Time, Health and Medicine*. London: Sage, pp. 1–30.

Fraser, G. 1995. 'Modern Bodies, Modern Minds: Midwifery and Reproductive Change in an African American Community', in F. Ginsburg and R. Rapp (eds), *Conceiving the New World Order: The Global Politics of Reproduction*. Berkeley: University of California Press, pp. 42–58.

Gell, A. 1992. *The Anthropology of Time: Cultural Constructions of Temporal Maps and Images*. Oxford: Berg.

Heintz, M. 2005. 'Time and the Work Ethic in Post-Socialist Romania', in W. James and D. Mills (eds), *The Qualities of Time: Anthropological Approaches*. Oxford: Berg, pp. 171–83.

Hertz, R. 1960[1907]. *Death and the Right Hand*. London: Cohen and West.

Hsu, E. 2005. 'Time Inscribed in Space, and the Process of Diagnosis in African and Chinese Medical Practices', in W. James and D. Mills (eds), *The Qualities of Time: Anthropological Approaches*. Oxford: Berg, pp. 155–70.

James, W. and D. Mills. 2005. 'Introduction: From Representation to Action in the Flow of Time', in W. James and D. Mills (eds), *The Qualities of Time: Anthropological Approaches*. Oxford: Berg, pp. 1–15.

Jordan, B. 1993. *Birth in Four Cultures: A Cross-Cultural Investigation of Childbirth in Yucatan, Holland, Sweden and the United States*. Prospect Heights, IL: Waveland Press.

Kirkham, M. 1989. 'Midwives and Information-giving During Labour', in S. Robinson and A.M. Thompson (eds), *Midwives, Research and Childbirth, Volume 1*. London: Chapman and Hall, pp. 117–38.

Kleinman, A. and L. Eisenberg. 1981. *Culture, Illness, and Healing, Volume 1: The Relevance of Social Science for Medicine*. Dordrecht: Reidel.

Levi-Strauss, C. 1969. *The Elementary Structures of Kinship*. Boston: Beacon Press.

Loudon, I. 1992. *Death in Childbirth: An International Study of Maternal Care and Maternal Mortality*. Oxford: Clarendon Press.

Lowis, G. and P. McCaffery. 2005. 'Sociological Factors Affecting the Medicalization of Midwifery', in E. van Teijlingen (ed.) *Midwifery and the Medicalization of Childbirth: Comparative Perspectives*. New York: Nova Science Publishers, pp. 5–41.

Martin, E. 1989. *The Woman in the Body*. Milton Keynes: Open University Press.

Marx, K. 1906. *Capital: A Critique of Political Economy*. New York: Modern Library.

Mills, M. 2005. 'Living in Time's Shadow: Pollution, Purification and Fractured Temporalities in Buddhist Ladakh', in W. James and D. Mills (eds), *The Qualities of Time: Anthropological Approaches*. Oxford: Berg, pp. 349–66.

Parsons, T. 1951. *The Social System*. New York: Free Press.

Pizzini, F. 1992. 'Women's Time, Institutional Time', in R. Frankenberg (ed), *Time, Health and Medicine*. London: Sage, pp. 68–74.

Scheper-Hughes, N. and M. Lock. 1987. 'The Mindful Body: A Prolegomenon to Future Work in Medical Anthropology', *Medical Anthropology Quarterly* 1: 6–41.

Simonds, W. 2002. 'Watching the Clock: Keeping Time During Pregnancy, Birth, and Postpartum Experiences', *Social Science and Medicine* 55: 559–70.

Thomas, H. 1992. 'Time and the Cervix', in R. Frankenberg (ed), *Time, Health and Medicine*. London: Sage, pp. 56–67.

Thompson, E.P. 1967. 'Time, Work Discipline and Industrial Capitalism', *Past and Present* 38: 56–97.

Turner, B. 1987. *Medical Power and Social Knowledge*. London: Sage.

Van Gennep, A. 1960[1909]. *The Rites of Passage*. London: Routledge and Kegan Paul.

Van Hollen, C. 2003. *Birth on the Threshold: Childbirth and Modernity in South India*. Berkeley: University of California Press.

Weber, M. 1930. *The Protestant Ethic and the Spirit of Capitalism*. London: Allen and Unwin.

Young, A. 1982. 'The Anthropologies of Illness and Sickness', *Annual Review of Anthropology* 11: 257–85.

# Part II

## TIME AND CHILDBIRTH PRACTICES

CHAPTER 3

# COUNTING TIME IN PREGNANCY AND LABOUR

*Soo Downe and Fiona Dykes*

## Introduction

the knowledge of temporal organisation is not synonymous with the lived experience of time.

—Flaherty (1991: 76)

In this chapter we discuss some of the ways in which progress, duration and time are currently measured and monitored during pregnancy and labour. We recognize with many other writers in this area that the current approach to maternity care is located within the dominant paradigm of biomedicine and linearity. Our chapter uses a number of discourses, including feminism, consumerism, socio-economics and complexity theory. The critique we make of current authoritative approaches to time in the context of pregnancy and birth is based on a social constructivist interpretation (Boudourides 2003). The chapter commences with a brief summary of the concepts we will use. We then critically discuss three specific practices in current maternity care. These are the dating of and decisions about the normal duration of pregnancy; the measurement and representation of progress in labour; and the programme termed 'active management of labour'. In this last section, we use critical discourse analysis. This approach illuminates the ways in which discourse is both shaped by relations of power and ideology and has a constructive effect upon systems of knowledge in a dialectical relationship (Fairclough 1992).

## Some Relevant Concepts and Philosophies

We come to this chapter with a variety of ways of seeing the world. Both of us would accept a feminist interpretation that for some women childbirth management has been used oppressively. We also believe that one of the projects of researchers in this field should be to liberate women from such oppression, and to maximize the opportunities for them to be empowered and energized by their birth experiences. In this we agree with the interpretations of feminists such as Arms (1994), Oakley (1986) and Martin (1987). Indeed, one of us (SD) came to midwifery with the belief that childbirth can be a site of feminist praxis, or change, and that such change could make a profound difference to individual women and to society. However, this idealism is tempered with the recognition of the power of consumerism as a dominant ideology in modern society.

We see this second concept, consumerism, as a logical conclusion of late capitalism. It emphasizes the individual, and the quick fix – getting what is wanted as soon as it is wanted with as little effort and risk as possible. It is mirrored by a risk-averse, blame-prevalent social culture. We would argue that the spread of this ideology into childbirth (by professionals but also among some women), is epitomized by the emphasis on a view that pregnancy and birth can and should be as short as possible while insisting on guarantees of safety, on increased rates of epidural analgesia usage and caesarean section, and on litigation if and when things deviate from what was expected or demanded.

Our third way of seeing, through a socio-economic lens, allows us to re-examine the late capitalist modernist interpretation of time as a commodity, to be measured, costed and limited. Finally, the insights of complexity theory bring us to a critique of the assumption that time is chronological, linear and regular, and that progress in pregnancy and birth must therefore inevitably be understood in this way.

The underpinning perspective for our critique, social constructivism, allows us to conclude that there are multiple legitimate ways of seeing and doing society, science, and technology. Thus we argue that birth and time are socially constructed, unfixed and subject to debate. This view contrasts with late capitalist, modernist understandings of a world that can be controlled, that a particular kind of science has an answer for everything, and that most, if not all, products, resources, and possibly even values in society can be priced and traded. As we will discuss, a social constructivist interpretation acknowledges the current authority of this worldview, but does not privilege it as the only possible or best view.

## Concepts of Time

Flaherty has observed that 'the person learns the ordinary correspondence between experience and standard temporal units' (1991: 84). As a statement of universal fact, this is arguable. For example, Nurit Bird-

David (2004) noted that the Nayaka people of South India do not use a past-present distinction to construct time. It is, however, likely that the construct of time set out by Flaherty is one experienced by many people in industrial and post-industrial societies. Standardization of time units becomes more necessary when communications extend from small cohesive groups to regional, national and even international networks. Classically, the drive to standardize clock time was precipitated by, for example, the need to co-ordinate trains between different time jurisdictions in newly industrializing societies. The consequence of modern international communications and technical developments is an increasing search for standardization and precision, to the point where it has been noted of a particular atomic clock that: 'NIST-7, a caesium clock at the National Institute of Standards and Technology (NIST), is accurate to five parts in $10^{15}$... To put it another way, this clock will stay within one second of true time for 6 million years'.[1] Most clocks do not use a caesium base, but even standard atomic clocks, which are the basis of International Atomic time (TAI), are adjusted by one second (a 'leap second') most years. This is because TAI is based on stable astronomical observations, but the earth is slowing down in relation to these observations. The 'leap second' compensates for this variation.

While such precision is invaluable in certain contexts, it is meaningless at the individual, embodied level. From this human perspective, Friedman (1990: 14, 54) described six ways that the duration of time is subjectively understood:
1. engaging tasks make time pass more quickly,
2. more events lengthen the perception of duration,
3. aging accelerates the speed that time appears to pass,
4. a given interval seems longer if a judgment of the duration is anticipated
5. a duration seems longer if we are frustrated, waiting for a positive experience, waiting for a specific event, or in fear of imminent danger,
6. an interval seems longer if it is remembered in more detailed pieces and shorter if we think of it more simply

We will refer to a number of these parameters in our critique in this chapter.

## The Length of a Piece of String?
## The Dating and Timing of Pregnancy

Paradoxically, technological advances such as sophisticated dating tests and very early ultrasound scanning have differential effects on women's perception of time in pregnancy. In the case of dating tests, pregnancies can be diagnosed almost from the point of conception, and certainly some weeks before most women experience any embodied sensations that might indicate that their body is different. This changes the absolute length of time that a woman feels herself to be pregnant, as indicated by Friedman's fourth and fifth parameter of time perception above. On the other hand, Rothman (1994) has indicated that the possibilities and threats of sophisticated screening techniques, such as diagnostic ultrasound scanning and amniocentesis, can cause some women to put their embodied experiences of their pregnancy 'on hold' while they wait for the results of the tests: 'Nancy ... called her three weeks of waiting ... a period of "suspended animation" during which she was: "trying to delay the reality of the pregnancy to myself ... It was very difficult especially as the baby had started to move"' (Rothman 1994: 103).

In an insightful analysis of the role of time in pregnancy and labour in the U.S., Simonds (2002) noted themes of milestones, increments, duration, pace and surveillance, all of which have strong connotations of linearity. She observed that, in the textbooks she reviewed, pregnancy was based on a 'rewinding' of time. This observation is based on the Naegles rule, which constructs the length of a pregnancy by adding seven days to the assumed date of the last menstrual period, then counting backwards by three months (plus a year). Research in this area has challenged this rule, however. For example, retrospective analysis of the length of pregnancy in uncomplicated situations called for the addition of fifteen days for women having their first baby or ten days for those in subsequent pregnancies, after counting three months back (Mittendorf et al. 1990). The interesting factor here is that there is some attempt to differentiate between different groups of women, in contrast to the usual practice of applying Naegles rule universally.

While the practice of counting back may seem counterintuitive to the notion that time runs forward, the use of Naegles rule is at least based on an approximation, since, in cultures using the Gregorian calendar, months vary in length. This allows for some fuzziness around the progress of pregnancy. Despite the range of other calendars in common use across the world, the Gregorian calendar and hence Naegles rule are most commonly used in most maternity care systems in the world. However, in some places, such as Japan, there is a tradition of counting pregnancy in lunar months. Since there are ten missed cycles, pregnancy is deemed as being ten months long (Hashimoto 2006).

In contrast to the range of reckonings of calendar time across the world, ultrasonographic dating of pregnancy provides an illusion of precision and certainty that opens the door to management against rigid timelines. For example, Nguyen et al. (1999) conducted a study of a large series of women using two different measures, one ultrasonographic (fetal biparietal diameter, BPD, or the circumference of the fetal head), and the other the method of last menstrual period (LMP). The date of spontaneous labour for each woman in the study was predicted from various clinical models using these measures, with two assumptions: one, that the average length of pregnancy was 280 days; and the other, that it was 282 days. For both BPD and LMP, the gap between the predicted date of birth and the actual onset of labour was an average of around eight days, indicating the uncertainty around this area. However, despite this evidence of individual variability, the authors go on to recommend that a precise number of days (282) be added to the last menstrual period to calculate the expected date of labour for every woman. In addition, the measure for the efficacy of the calculation was based on so-called 'post-term' or 'pre-term' births. These were not differentiated by pathology, but by time. The authors showed that the number of post-term births fell from 7.9 per cent to 5.2 per cent and the number of pre-term births rose from 3.9 per cent to 4.4 per cent using this new way of dating pregnancy. In this construction, the timing of pregnancy becomes the measure of abnormality, irrespective of actual or even potential pathology. This work illustrates that the persistent search for the certainty of precise clock time has the potential to create the kind of maternity care women experience, irrespective of their individual physiology, family history and actual obstetric risk factors.

Simonds has noted that, 'hospital staff manage bodies in time and space by regulating them in ways that contribute to conducting the business of the institution efficiently, as determined by the "experts" who run it' (2002: 564). While this is insightful in the light of the discussion above, it locates the power to manage with hospital staff alone and ignores the impact of such ways of managing on the expectations and demands of women. The notion of technological certainty raises expectations in women that their pregnancy will end within a narrow time band. This has potential knock on effects for the embodied sense of time in the waiting pregnant woman, as discussed earlier. While in most cases, risk management regimes mean that professionals feel obliged to persuade women to accept induction of labour once it is 'over time', the consequence in some cases may also be a consumerist pressure on maternity staff to induce labour when this narrow time band is approaching, or has passed.[2] In this way, the clock-time view of pregnancy becomes reinforced, as women do indeed have their babies within an ever-narrower time frame that is closely aligned to the population based 'norms' of the technology of surveillance.

Indeed, even in settings that are set up to offer alternatives, the clock-based imperative can intervene. As an example, in Annandale's (1988) study of a birth centre, she observed what she terms 'ironic intervention' by the midwives. This term denotes interventions used by midwives to pre-empt transfer of women, and to therefore avoid intervention in the hospital. In one particular case, Annandale noted that after the second half of 1983 there was a steep rise in births occurring between forty weeks and one day, and forty-two weeks gestation. This correlated directly with an edict from the central hospital that women over forty-two weeks gestation must be transferred automatically to obstetric care. In using interventions such as castor oil to try to ensure that women went into labour prior to the cut-off date for transfer, these midwives were unwittingly colluding in the creation of a population of women for whom the 'norm' was to go into labour prior to forty-two weeks gestation. If this is replicated in other areas of childbirth, future epidemiological studies are likely to conclude that the population norms for the timing of pregnancy and labour are narrower than they would have been if no intervention had taken place. The pregnancies and labours of women outside these 'norms' are then increasingly seen as pathological and more likely to be subject to the technocratic gaze. This process of reification also sets women up to expect a labour that is predictable and controlled, an example of Flaherty's (1991) learned correspondence between experience and standard temporal units.

In contrast to the certainties of pregnancy standardized by clock time, a few authors have called for women to be given a range of dates for the end of their pregnancy. For example, Saunders and Paterson claimed that 'much anxiety would be alleviated if a range of dates (38–42 weeks) was substituted for a specific date of delivery' (1991: 600). However, in the light of the increasing insistence on precision over the last fifteen years in maternity care, this plea seems to have been ignored, both by practitioners and by service users. This is possibly not surprising as contemporary culture is characterized by a desire for certainty, and 'control' is a mantra of modern childbirth. Being given a precise date not only allows hospital staff to plan for the likely number of births within a narrow time band, but also allows women and families to arrange for maternity leave and for support systems around the given due date. Even though these plans are of necessity imprecise (since most babies are not born on their due date) they give the illusion of control and are thus preferred to the (more accurate) uncertainty that is offered by a range of possible due dates.

## 'It Takes as Long as It Takes': Labour and Birth Time

> the phase of maximum slope is a good measure of the overall efficiency of the 'machine' with which we are dealing.
>
> —Friedman (1978: 34)

The industrial baby-as-product model of birth has been widely described and critiqued (Oakley 1986; Martin 1987; Arms 1994; Wagner 1994). We discuss this in more detail in the next section. More recently, Davis Floyd (2003) has described a late-modernist technocratic turn in ways of doing birth. Both interpretations view birthing bodies as machines or technologies that operate in narrow bands of acceptability, and that are prone to error, but correctable. In this paradigm, birth is only controllable by imposing strict time limits.

The quote used in the title of this section ('it takes as long as it takes') is in stark contrast to the industrialized, technocratic model of birth. It comes from one of the respondents to an interview study of midwives in New Zealand (Crabtree 2008: 108). Most of the midwives cited in the study appeared to be operating in the spirit of this laissez-faire attitude.

As another respondent stated: 'I think I can last a lot longer [than other midwives] because I have seen so many people achieve, with time. I don't think that there is a rush in labour. I don't have a Friedman's curve timeline in my practice' (Crabtree 2008: 100). This quote illustrates the point that midwives are not all united in their conception of labour time as a fluid concept, even when the socio-economic circumstances suggest that they could be. Second, this midwife is recognizing the individuality of time and achievement in labour, namely that there are not hard and fast and linear rules.

The standard 'normal' length of labour became more narrowly defined through the twentieth century. As *Myles' Midwifery*, a classic midwifery textbook, stated in its thirteenth edition, 'the active phase [of labour] proceeds at a rate of 1–1.5 centimetres per hour and a rate of 1 centimetre per hour ... is commonly accepted as the cut off between normal and abnormal labour (Cassidy 1999: 408). There is no obvious scientific basis for this rule, and its parameters have changed over time (Downe 2004). The accepted 'normal' rate of approximately one centimetre per hour is commonly believed to be based on the work of Friedman (1955). However, Friedman's classic study actually indicated that labour progresses at an average rate of 1.2 centimetres per hour, and that this progression is not linear, but sigmoid (see Figure 3.1).

Friedman's findings were initially derived from a study of 100 American women having their first baby in one New York hospital, based on frequent (in some women half-hourly) rectal examination in labour. This work

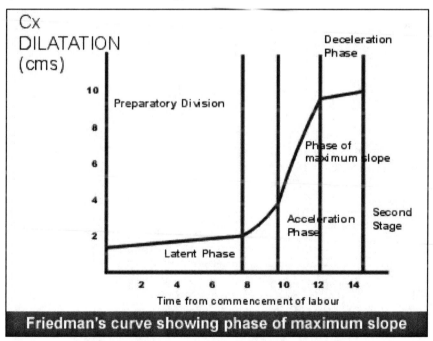

Figure 3.1: Friedman's Curve.
Source: http://uk.geocities.com/drsandhyasaharan/images/fried.jpg

was then translated into a graphic representation of the pattern of cervical dilation, the so-called 'Friedman's curve'. Friedman acknowledged that other researchers had presented cervical dilation graphically before he examined this area (Friedman 1978). However, the labour pattern described in his 1955 publication, and subsequently in an expanded work published nearly twenty-five years later (Friedman 1978), has become almost iconic in maternity care across the world. The fact that a number of authors since Friedman have described the same pattern across different populations, their results latterly based on vaginal examination, gives it some credibility. It should be noted, however, that digital examination of the cervix stimulates prostaglandin release. Prostaglandin is a hormone that has an effect on the uterine muscle, making it more likely to contract. This suggests that the commonality in the findings of studies looking at the 'average' length of labour may be at least in part an artefact of the systems of maternity care that are based on measuring progress using regular cervical examination. In other words, while it is likely that many women would labour within the parameters of Friedman's curve if left alone, there is probably a substantial minority who would not do so, but who would still be well and healthy. This is the group for whom labour is deemed potentially pathological in cultural settings where time and standard measurements are the arbiter of normal and abnormal.

As Lavender (2003) has noted, the graphical measure of labour produced by Friedman is properly termed a cervicograph, since it was only concerned with cervical dilation and station of the fetal head in relation to the maternal ischial spines (a part of the pelvic bones). Indeed, Friedman explicitly stated that these were the only two parameters that made any difference in labour (1978: 17). As can be seen from figure 3.2, the classic labour partogram (a graphical measure of parturition), in contrast, records a large number of factors, including the frequency and strength of contractions, a measure that was explicitly rejected by Friedman as useful in this context.

The alert and action lines noted in figure 3.2 are based on the work of Philpott (1972) and Philpott and Castle (1972a, 1972b). They were originally designed as a guide for rural carers in Rhodesia (now Zimbabwe) to when they should be considering, and then taking action on, an abnormally slow labour. This was specifically tailored to the local context, where transfer of women from rural settings may take hours, or even days. As for the work of Nguyen et al. (1999) discussed above, the definition of abnormal was as a statistical rather than a directly pathological concept. The time attributed to abnormally slow labours was extrapolated from the 10 per cent of women with the slowest labours in their particular population. Based on the work in Rhodesia, Studd and Philpott (1972) devised a set of labour stencils to describe the expected pattern of labour for women depending on their dilation at the time of admission to hospital. This was then exported to Western hospitals for women across the risk range. In

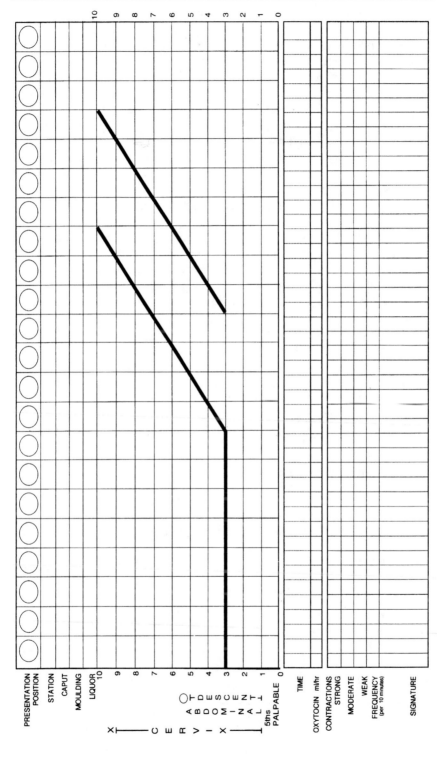

Figure 3.2: Partogramme with alert and action lines.
Source: Liverpool Women's Hospital, 2009. Reproduced with permission.

this case, where transfer times are minutes or hours at the most and, indeed, where most women are in hospital for their births, the benefits of early assessment of slow labour are much less obvious. Despite this, the use of one version or another of the partogram based on the work of Friedman and the researchers who followed him is now nearly universal in most intrapartum maternity care settings. It has been exported back to rural communities in Africa and Asia, and a recent study by the Worl Health Organization has found benefits for women and babies when it is used in these circumstances (WHO 1994). However, while the placement of the action line has been assessed in a Western hospital-based setting (Lavender 2003), the risks and benefits of the partogram itself have not been assessed in this context.

In a move that has paralleled many developments in human health, and as we noted above in relation to the dating of pregnancy, the partogram provides a neat rationale for an increasingly precise definition of how long labour should be, and how the intervals of time within it should be divided up. The overall accepted norms for the length of the second stage of labour have become more closely defined over the years, starting with an acceptance that no one can predict how long a labour can be in individual cases, and progressing with increasingly shorter definitions of what is normal. For example Berkeley, Fairbairn and White suggested that: 'labour in primiparae lasts on average about 15 to 20 hours. In multiparae, eight to ten hours can be taken as the average time.... It is foolish to attempt to prophesy more than approximately how long labour will last in any given case' (1931: 273).

By 1958, Myles's textbook for midwives appears to be rather more certain and her statement uses the language of normality as well as that of the average:

> the multiparous woman may have a second stage of 15 minutes or less and one hour is advocated as the limit of normal. The duration of the third stage is between 10 and 20 minutes, and one hour is recommended as the limit of what could be considered normal. The consensus of opinion is that during the past twenty years the duration of labour has been shorter than previously, probably due to the greater use of relaxation and sedation.... A considerable number of primiparae have labours of under 12 hours, and by far the greater numbers of multiparae have labours of 6 to 8 hours and in many cases less than 6 hours. (1958: 228)

These figures are reinforced in a subsequent textbook by Myles (1975: 211).

|              | First stage | Second stage | Third stage | total |
|--------------|-------------|--------------|-------------|-------|
| Primigravida | 11          | 3/4          | 1/4         | 12    |
| Multipara    | 6 1/2       | 1/4          | 1/4         | 7     |

This later edition also stated that the duration of labour had decreased in the last twenty years and added oxytocin (used to speed up labour contractions) to the list of elements considered to have contributed to the change (Downe 2004).

The division of this timescale into shorter and shorter and more and more apparently precise units does not allow for individual variation. Indeed, since it is defined on a linear scale, it does not even take account of the sigmoid nature of Friedman's curve, where cervical dilation is slower at the beginning and the end of labour. Even where research has been undertaken explicitly to challenge the hegemony of current definitions of normal labour progress, the conclusion is framed in linear terms. For example, Albers and her colleagues undertook a series of studies of low risk spontaneously labouring women, and concluded that labour was longer on average than in the standard textbook definitions (Albers, Schiff and Gorwoda 1996; Albers 1999). However, the findings were still framed in centimetres dilation per hour for the population studied.

In contrast, some midwives appear to take a view that the processes and progresses of labour fit better with an individualized complexity model rather than one based on simplified population averages. Complexity theory is based on the observation that many physical processes are not linear and predictable but are uncertain and emergent, built on connectivity and on what has gone before, or 'sensitivity to initial conditions' (Sweeney and Griffiths 2002; see also Chapter 4, this volume). Complexity theory would argue that progress in labour is a product of the connectivity between a variety of hormonal, physiological and psychological cues between mother and baby and mother and the external environment. If the technocratic model as captured in the output of cervical dilation can be said to see bodies as machines, the physiological model based on concepts of connectivity can be equated to seeing labour as music, based on the observations of Husserl:

> What Husserl discovers is a whole new dimension of temporal experience. In addition to the experience of the immediately present note, and the recall in the present of a memory of a past note, there is also what Husserl calls a retention of past notes – a kind of implicit presence of the just past. Although a sound may have just ended, there is a continuing retention of it in the present that gradually fades.[3]

In this context, time is a more fluid phenomenon than current conventional science would accept.

Critical analysts in maternity care tend to polarize the dominant professional groups paradigmatically. Obstetricians are seen as the protagonists for interventionist approaches in the name of safety, and midwives are identified with a 'wait and see' attitude. However, as Arney (1982) has illustrated, adopting a Foucauldian perspective, this division is far from straightforward. The way of doing birth that is dictated by technocratic, mechanistic, risk averse, consumerist and above all surveillance-oriented societies can dictate the actions of any professional group operating within it. In this context, midwives and women as well as obstetricians learn to view birth as a linear process, with clear beginnings and ends. We discuss the impact of notions of beginning and ending more fully in the next section.

The example of the 'early pushing urge' is useful in illustrating the clash between embodied time in labour dictated by physical sensations, and clock time dictated by cervical dilation. In most modern maternity care settings, the end of the first stage of labour is defined by a cervical dilation said to be ten centimetres. As in the measurement of pregnancy, this precision is an artefact, since women and babies differ in their physiology, and the artifice of precision is used to justify the insistence on conformity. Women are not supposed to actively push in labour until this cervical dilation has been confirmed by an expert (doctor or midwife). Their own sensations, which are usually very powerful, are discounted in this decision. A recent survey has suggested that at least 20 per cent of women may be physiologically in the pushing stage of labour before their cervical dilation is complete (Downe et al. 2008). In these cases, the complex adaptive processes of labour are out of step with simplistic, linear cervical time.

Women feeling a strong urge to bear down who are found not to be 'fully dilated' are told 'it is not yet time to push' and are urged to find ways of overcoming the pushing reflex until deemed 'ready'. A systematic review and two surveys of midwife practice have suggested that there is no good evidence for this approach to the early pushing urge (Downe et al. 2008). Bergstrom et al. (1997) gave a graphic description of the distress this caused one woman. Indeed, in this case, as in other case histories, the woman actually lost the sensation to push when deemed 'ready' and, ultimately, the baby has to be delivered using forceps. However, despite the very powerful out-of-time sensations women experience during this so-called transitional phase, many still acquiesce with the clock-time concept of labour, as they are taught to do in antenatal sessions. For example, in Anderson's study of women's accounts of their relationships with their midwife while they were in labour, 'all of the women had attended antenatal classes ... and had clear expectations that the midwife would tell them when they could push. If this did not happen, they were left in limbo and not in control, not knowing whether or not to push' (2000: 111). In the only qualitative empirical study we could locate that was specifically focused on women's perceptions of time during labour, Beck reported the findings of phenomenological interviews with seven women on their second postpartum day. Her detailed analysis of the data led her to conclude that: 'the paradoxical and fluctuating nature of women's temporal experiences during the delivery process emerged. While absorbed in labor, time seemed endless, but ... women experienced amazement and disbelief in how much time had actually passed when they validated their subjective time with the clock in their labor rooms' (Beck 1994: 245). This observation indicates that at the subjective extreme of labour, time is a fluid experience. In contrast, the next section explores the origin and operation of an extreme example of a clock-time model.

## Active Management of Labour

To further illustrate the deeply embedded nature of linear time within Western obstetric approaches, we draw upon a seminal text, *Active Management of Labour* (O'Driscoll, Meagher and Boylan 1993). This text had a major influence upon medical practice with regard to birthing women in the latter parts of the twentieth century. There were three editions of the book – in 1980, 1986 and 1993 – with only minor changes between them; the third edition is referred to here. Sales of the book extended to North America, Australasia, Europe and parts of the developing world. Translation into Spanish facilitated a more global uptake. The book was authored by three consultant obstetricians who were based at a large obstetric unit in Ireland. Although over thirty years have elapsed since the book was first published, the application of its principles is still evident to varying degrees in contemporary obstetric units across the globe. Although the rise of consumerism has seen the language of texts change, active management continues to dominate hospital practice, via protocol-based care, in much of the world, regardless of local context.

The 'active management of labour' concept was initiated by the authors and 'evolved' over twenty-five years (O'Driscoll, Meagher and Boylan 1993: 13). In essence, this approach involves: obstetric involvement in all labouring women's 'management'; emphasis upon 'correct diagnosis' of a woman's labour; precise timing and measurement of the progress of labour; early augmentation of 'inefficient uterine action' according to a strict medical protocol; antenatal education based on the protocols; providing each woman with continuous personal support; and rigorous audit of events and outcomes. The book is divided into three sections. Section I refers to 'management' and 'roles' in relation to staff and labouring women. Section II consists of twenty-eight partograms, some of which illustrate particular problems, with an accompanying description on the adjacent page. Finally, Section III consists of pages of audit data taken from the hospital, which the authors asserted speak for themselves with regard to their approach to labour.

The use the term 'active' by O'Driscoll, Meagher and Boylan has been heavily critiqued as applicable to the obstetrician while the woman appears to be relatively passive (Kirkham 1983; Oakley 1986; Martin 1987; Schwarz 1990; Wagner 1994; Goer 1995; Tew 1995; Murphy-Lawless 1998). For example, Oakley argued that, 'the notion of an actively managed labour – or the vision of all labours actively managed – summarises the uncertainty of professional attenders of childbearing women in the face of a "natural" process whose unknown laws undermine their would-be omniscient human expertise' (1986: 205).

### The Uterine Machine.

Although O'Driscoll, Meagher and Boylan (1993) emphasized the importance of the woman, their language was suggestive of a clear conceptual compartmentalizing of the mind and parts of the body that has become a characteristic of biomedicine. There is repeated reference to the efficiency of the uterus with statements such as 'efficient uterine action is the key to normality' (O'Driscoll, Meagher and Boylan 1993: 16). This reflects the mechanistic theme noted above, and seen in obstetric texts such as Llewellyn-Jones (1982), in which the metaphors of the uterus as machine, providing the power, the fetus as passenger and the genital tract as the passages, are commonly depicted (1982: 77–104). Indeed, it is the need to ensure that the uterus, represented as a machine and prone to failure, has enough power to drive the somewhat reluctant fetus though the often unaccommodating reproductive tract that seems to form the basis for the active management of labour. As we have noted, Davis-Floyd referred to these metaphors as representing the characteristics of the technocratic paradigm of birth in which the 'female body is viewed as an abnormal, unpredictable and inherently defective machine,' particularly prone to 'serious malfunction' in pregnancy and birth (1994: 1127).

Notions of the uterus as a semi-independent machine continue through chapter headings in O'Driscoll, Meagher and Boylan's text, such as 'cephalopelvic disproportion' and 'cervix in labour' with the 'owner', the woman, missing from the titles. This linguistic separation and invisibility of women is highlighted in a number of other critical studies (Oakley 1986; Martin 1987; Petchesky 1987; Franklin 1991; Duden 1993; Dykes 2002, 2005, 2006).

## A Beginning and an End

A key emphasis of the text appears to be a focus on an active and dynamic approach to labour, based on ensuring efficient uterine action. Active and dynamic here appear to refer to the obstetrician. The authors challenged some obstetric practices in other centres by claiming that there is a policy of watchful expectancy during the 'tedious' first stage of labour, followed by a sudden resort to intervention for the birth (O'Driscoll, Meagher and Boylan 1993: 13). The word tedious appears to convey the idea of slowness in labour as boring and a nuisance.

Efficiency, in an industrial sense, relates to production against recommended measurements of linear time. As we have noted above, this connects with the western linear assumption that 'every event or phenomenon will have both a beginning and an end' (Helman 2000: 37). These notions were firmly embedded in O'Driscoll, Meagher and Boylan's emphasis upon determining, with precision, the point at which a woman's labour commences: 'As with every syndrome in clinical medicine, labour too must begin with a correct diagnosis' (1993: 43). The authors state that labour is defined 'as the number of hours a woman spends in the delivery unit, from the point of her admission until the time her baby is born', adding that 'no allowance is made for time spent in labour at home' (1993: 32). The latter statement appears to lie in contradiction to the precision emphasized in relation to diagnosis of the onset of labour.

Not only is the beginning of labour emphasized but also the end in that O'Driscoll, Meagher and Boylan state that every expectant mother is given a firm assurance that her labour will be completed within twelve hours. Therefore, if the woman's labour does not progress at the prescribed rate two interventions are available to enhance her uterine efficiency, amniotomy and the administration of oxytocin. The authors were so clear that labour should not exceed twelve hours that, as Oakley (1986) pointed out, their graphs made no provision for a labour that lasts beyond this period of time. The role to determine the end of labour rested firmly with a senior doctor: 'The senior registrar, in consultation always with the midwife in charge, decides when to terminate labour and also chooses the method of delivery except in the case of a caesarean section that must be referred to the consultant' (O'Driscoll, Meagher and Boylan 1993: 97).

Thus time is defined by the institution. This reconfiguring of time resonates with Simonds's assertion that in obstetric management, 'the notion that obstetrics can – and should – work against the bodily clock that stops too soon or continues too long has gained credence (2002: 563), a point that she follows up by noting that 'in obstetrics, time signifies the danger unmanaged women's bodies represent. Now women are bound by the clock rather than leather straps' (2002: 568).

### Making Orderly Progress

O'Driscoll, Meagher and Boylan (1993) place enormous emphasis upon the use of the partogram. As in other aspects of biomedicine, for example cancer treatment (Stacey 1997), the partogram enables emphasis to be placed upon concreteness, visibility, physicality and progress, presenting the phenomenon, in this case 'labour', as statistical, quantifiable and observable. The centrality of the partogram in many maternity units worldwide, along with fetal heart monitoring, contributes to the idea that a woman's labour is best represented as progress on a graph rather than being something that can be monitored through the feedback provided by her own experience. Duden (1993) highlighted this dependency upon visual data, claiming that it overrides sensations experienced by the woman.

The partogram section of O'Driscoll, Meagher and Boylan's text illustrates the display of data measurements relating to the woman's progress, in particular dilatation of the woman's cervix against time. The authors state that this enables the clinician to view labour as a 'concrete rather than an abstract pursuit' (O'Driscoll, Meagher and Boylan 1993: 18). Concreteness, in this context, clearly refers to the clinician's perspective. As the acceleration of 'slow' labour using synthetic oxytocin is a fundamental principle of this text, an entire chapter entitled 'oxytocin in labour' is devoted to the subject. In a section entitled 'explicit rules' the authors clarify the exact quantities of oxytocin per minute that should be administered to the woman with the rate increasing every fifteen minutes until a specified rate in drops per minute is reached. The authors' adherence to the merits of oxytocin is related to their assumption that uterine damage and cerebral injury to the baby are more likely to be the result of the use of forceps to provide traction and less likely to be related to the strength of uterine contractions. Thus, they appear to advocate a stoking up of the uterine machine which thereby pushes the passenger along rather than the option of pulling them out. While this would seem to be a logical position to take, and in most cases oxytocin is a relatively safe drug, there are potential problems with excessive use. According to the data sheet of one commercial company, these include: commonly, maternal nausea and vomiting, and fast or slow maternal heartbeat (tachycardia and bradycardia); less commonly, uterine hyper-stimulation; in rare cases, fetal distress, placental abruption and amniotic fluid embolism; and, very rarely, maternal or fetal death.[4]

So, as Simonds argued, obstetrics manages pregnancy and birth by 'institutionalising rigid time standards, carving procreative time up into increasingly fragmented units' (2002: 559–60). In this way, as Jordan (1997) described, privileged techno-medical knowledge, in this case related to time, tends to supersede and delegitimize other sources of knowledge generated from a woman's own bodily experiences.

### Monitoring Progress

The biomedical view that childbirth is potentially hazardous unless proven otherwise is clearly evident in statements such as 'ironically, it is in completely normal women that most of the problems in labour arise' (O'Driscoll, Meagher and Boylan 1993: 15). It is also illustrated in the descriptions of the partograms in Section II: 'these visual records or partographs, illustrate better than words ever can, the problems that arise in the course of everyday practice' (1993: 18). In view of the purported risks the close medical surveillance of all women is advocated with the delivery suite designated as an intensive care area. In this setting women's labour progress may be reviewed regularly, day and night. This close surveillance is clarified in the statement that the obstetrician should become involved directly with all women including 'perfectly normal women,' from the point of admission in labour (1993: 15). Thus the obstetricians appear to function, as Schwarz asserts, as chief 'engineers ... in charge of the smooth running of all systems' (1990: 35). The obstetrician may then introduce a notion of 'certainty' through monitoring against time. As Thomas stated:

> Time provides not only a way of describing the distribution of events but also a basis for interpretations and explanations. In reproduction it provides a way of distinguishing between the normal and the abnormal, between the abnormal and the pathological. More generally this can be seen as a differentiation of the ordered from the disordered, the orderly from the disorderly. (1992: 65)

Having established the consultant obstetrician as in ultimate charge, O'Driscoll, Meagher and Boylan defined the essential hierarchical structure that should be present on the delivery unit to optimize surveillance and management of labour. Military language is utilized to define the senior midwife's role. The authors stated that a 'chain of command must be sharply defined', with one 'nurse/midwife' in charge of the delivery unit who is dressed in 'distinctive uniform that can be recognized instantly by all concerned,' so that 'everyone who works in the unit should be subject to her immediate authority. There is no more room for divided responsibility in a delivery unit than there is aboard a ship at sea' (O'Driscoll, Meagher and Boylan 1993: 96). This person is given the responsibility to 'confirm or reject the diagnosis of labour in every case admitted', regularly measure the woman's cervical dilatation and decide when to accelerate labour (1993: 100). The hierarchical theme is pursued when the authors refer to a 'Mastership system', with the consultant obstetricians clearly in charge and responsible for the outcome of every woman's labour, so that 'subordinate staff' on the 'shop floor' are not anxious about becoming a 'scapegoat' if a mishap should occur (1993: 97).

Despite the explicit reference to the consultant obstetrician as being in charge, surveillance as a pervasive philosophy does not simply involve the senior obstetrician supervising the mother and her baby. As Arney stated, 'monitoring is the new order of obstetrical control to which not only women and their pregnancies are subject but to which obstetrical personnel themselves are subject' (1982: 102). Thus a system has been generated which insidiously encourages conformity to the techno-medical norms through a complex and multifaceted form of surveillance, including norms of time.

### Training for Progress

Goer pursued the military metaphor in relation to 'active management of labour' arguing that the mother appears to be relegated to the rank of 'private' (1995: 85). As 'private', the woman is required to understand notions of 'progress'. To equip her with this knowledge, O'Driscoll, Meagher and Boylan advocate the use of the partograms, displayed in Section II, as the 'focus of antenatal preparation for labour' (1993: 18). Later, in a chapter entitled 'antenatal preparation', it is stated that 'everyone is given a copy of the official partograph to take away for further study and all are expected to be familiar with the regimen when admitted eventually in labour' (1993: 78). This is an example of training mothers to view their labours as linear, time bounded events to be closely, medically monitored and managed according to a graphical display of data.

Women were not expected to be 'privates' alone. O'Driscoll, Meagher and Boylan promised that a 'personal nurse' will be in attendance at all times (1993: 94). The offer of a package of relative certainty, to include the pledge of a highly consistent approach, continuous one-to-one professional support and a labour of less than twelve hours in duration, may appeal to women in cultures in which control, certainty and linear time are the norms. The appeal of certainty invoked by technology was illustrated in a Western setting by Davis-Floyd (2003). She interviewed one hundred American, middle class women about their perceptions of birth and reported that almost half of the women saw modern birth facilities as empowering them to control an otherwise unpredictable biological experience. These women demonstrated 'cognitive ease' with the techno-medical model and had come to see technology as normal and beneficial (Davis-Floyd 2003: 239). This, as she argued, highlights the strong socializing power of the hegemonic techno-medical model.

### Contrasting Times

The analysis of O'Driscoll, Meagher and Boylan's (1993) text highlights some key issues in relation to the biomedical model with its focus upon linear time and orderly progression throughout women's labour and births. It reflects the mechanistic approach to the human body reflective of the technological era. As Fox (1989) argued, by substituting a male model of productivity for the archetype of the creative and transformative mother, birth has been reconfigured as a mechanical act and a time-bounded process.

Simonds (2002) contrasted obstetric, productive notions of time with those seen in Ina May Gaskin's well-known text, *Spiritual Midwifery* (1990). She asserted that *Spiritual Midwifery* exemplifies an 'essentialist orthodoxy wholly antithetical to medicalised environments and, thus, to prevalent cultural ideology about procreative events' (Simonds 2002: 569). Gaskin's text focuses upon women being active in time, with the focus moving clearly away from any notion of time constraint. Simonds stated in relation to Gaskin's text:

> Time is not something to be rationed; not a scarce commodity that procreating women waste; not a route toward measuring pathology nor, in itself, an indication of pathology; not a series of obstacles against which women's performance must be measured; and not a means of industrialising the labour process. Time just is. Birthing women can take their time, rather than have it taken from them. (2002: 569)

## Discussion

The synchronization of clock time for trains across England, and, indeed, between countries, did not make the time more 'real', just more transmittable. In the case of birth, arguably the main rationale for a linear-time-based orthodoxy is that it allows information to be transmitted between carers (or, indeed, to lawyers) based on a certain set of standardized data, such as days of pregnancy, centimetres of cervical dilation or length of time in labour. Arney's (1982) Foucauldian interpretation of the professional projects of both midwifery and obstetrics indicates that the need for surveillance in birth is a more powerful predictor of the use of techniques of control and management than is the project of these specific professional groups. In a technological, consumerist, surveillance-oriented society that believes in certainty and the possibility and importance of precision, time matters when information needs to be externalized, or when a practitioner wants to be believed or taken seriously.

Flaherty referred to the observation of Csikszentmibalyi that 'people who enjoy what they are doing enter into a state of "flow", they concentrate their attention on a limited stimulus field, forget personal problems, lose their sense of time and of themselves, feel competent and in control, and have a sense of harmony and union with their surroundings' (cited in Flaherty 1991: 76–7). Csikszentmibalyi (1975) hypothesized that this sense of flow is brought about by a tenuous, situated balance between one's skills and one's challenges. In contrast, Flaherty's (1991) study was focused on the perception amid students and university staff of time going slowly. He used the term 'synchronicity' for when individuals are more or less in tune with the general population's experience of time. Interestingly, for a discussion of pregnancy and childbirth, the term he used for the moments when time is perceived to be passing slowly was Goffman's word 'engrossment'. In Flaherty's analysis, these moments were either when situations entailed great urgency and affective arousal, or when people were experiencing apparently 'empty' times, which 'turn[ed] out to be quite full ... of perceptions, feelings, thoughts' (1991: 80). In both cases of engrossed time, 'lived time is perceived to swell as standard units of time become laden with the thick experience of subjective stimulus complexity' (1991: 82).

It can be hypothesized that the normally pregnant or spontaneously labouring woman is lost in the timeless harmony of Csikszentmibalyi's 'flow'. However, if the onlookers, including partners and professionals, are experiencing Flaherty's 'empty time', this will be exacerbated by the clock, resulting in a felt need to 'do' something, especially where time is seen as a precious commodity, not to be wasted. The 'doing' may well be seen to be intrusive and unnecessary by the pregnant or labouring woman engrossed in affectively aroused, embodied time. On the other hand, where women accept and demand precision, certainty and a quick fix, it is likely they will seek an early end to their pregnancy and labour. In this case (especially where epidural analgesia is used in labour) it is possible that the drag of 'empty thick time' will be felt by the woman as well as the staff, and the offer of intervention to 'speed things up' may not only be acceptable, but demanded (Davis-Floyd 2003). In this way, birth is reified as a technical-industrial process leading to a consumerist product.

In contrast, those women and practitioners who see pregnancy and birth as a rite of passage perceive it as a liminal, timeless, transformational moment that is both unique and a part of normal life time. As Grimes noted, 'A rite of passage is like a domestic threshold or a frontier between two nations. Such places are "neither here nor there" but rather "betwixt and between"' (2000: 6). For women who take this approach, 'it may well be that ... losing track of time is a very helpful coping strategy' (Anderson 2000: 98). As we have demonstrated, this latter group of women and practitioners may be swimming against the tide of current authoritative approaches to childbirth.

# Conclusions

We have critically discussed three aspects of current maternity care, namely the dating of and decisions about the normal duration of pregnancy; the measurement and representation of progress in labour; and the 'active management of labour'. We have noted the ways in which discourse is shaped by power and ideology, and how, in turn, authoritative discourses reify cultural knowledge. This has led to the iconic status of measures of time in maternity that offer the promise of time limitation and control, such as partograms based on Friedman's curve. While the use of such measures in some circumstances is undeniably life saving, critical analysis of their widespread usage leads us to conclude that they have come to serve a global economy where the dominant ideology is consumerism, where illusions of certainty and control over life events are valued, and where time is money. Those who value the uncertain, liminal, 'becoming' aspects of pregnancy and childbirth will need to find persuasive arguments to counter the current hegemony of risk-averse, time-limited childbirth.

# Notes

1. Source of information: http://whyfiles.org/078time/2.txt. Retrieved [1st December 2008].
2. Poststructuralist sociologists such as Haraway (1991) and Annandale and Clark (1996) have argued that in a technocratic society it should not be surprising if women begin to expect or even demand technological interventions as part of a general consumerist attitude. Additionally, a number of obstetricians have argued that recent rises in obstetric-intervention rates are largely a result of consumerist pressure. However, there is little substantive evidence that consumerist pressure is having such an effect and studies largely confirm that women's choices are constrained by structures of power and have only limited impact on the delivery of maternity services (Gamble et al. 2007; McCourt 2006).
3. Quoted from: McFarlane, T.J. 1998. 'The Nature of Time'. Retrieved [1st December 2008] from: http://www.integralscience.org/abouttime.html.
4. Source of information: Novartis New Zealand. 2006. 'Syntocinon®: Information for Health Professionals'. Retrieved 4 July 2006 from: http://www.medsafe.govt.nz/Profs/Datasheet/s/syntocinoninj.htm.

# References

Albers, L.L. 1999. 'The Duration of Labor in Healthy Women', *Journal of Perinatology* 19(2): 114–19.

Albers, L.L., M. Schiff and J.G. Gorwoda. 1996. 'The Length of Active Labor in Normal Pregnancies', *Obstetrics and Gynecology* 87(3): 355–59.

Anderson, T. 2000. 'Feeling Safe Enough to Let Go', in M. Kirkham (ed.) *The Midwife–mother Relationship*. Basingstoke: Macmillan, pp. 92–119.

Annandale, E. 1988. 'How Midwives Accomplish Natural Birth: Managing Risk and Balancing Expectations', *Social Problems* 35: 95–110.

Annandale, E. and J. Clark. 1996. 'What is Gender? Feminist Theory and the Sociology of Human Reproduction', *Sociology of Health and Illness* 18(1): 17–44.

Arms, S. 1994. *Immaculate Deception: Myth Magic and Birth,* 2nd edn. Berkeley, CA: Celestial Arts.

Arney, W.R. 1982. *Power and the Profession of Obstetrics.* Chicago: University of Chicago Press.

Beck, C. 1994. 'Women's Temporal Experiences During the Delivery Process: A Phenomenological Study', *International Journal of Nursing Studies* 31(3): 245–52.

Bergstrom, L., et al. 1997. 'I Gotta Push. Please Let Me Push! Social Interactions During the Change from First to Second Stage Labor', *Birth* 24(3): 173–80.

Berkeley, C. and Fairburn White. 1931. *Midwifery by Ten Teachers,* 4th edn. London: Edward Arnold.

Bird-David, N. 2004. 'No Past, No Present: A Critical-Nayaka Perspective on Cultural Remembering', *American Ethnologist* 31(3): 406–21.

Boudourides, M.A. 2003. 'Constructivism, Education, Science, and Technology' *Canadian Journal of Learning and Technology* 29(3). Retrieved [1st December 2008] from: http://www.tonybates.ca/2008/07/21/constructivism-education-science-and-technology.

Cassidy, P. 1999. 'The First Stage of Labour: Physiology and Early Care', in V.R. Bennett and L.K. Brown (eds), *Myles' Textbook for Midwives.* Edinburgh: Churchill Livingstone, pp. 391–410.

Crabtree, S. 2008. 'Midwives Constructing "Normal" Birth', in S. Downe (ed.) *Normal Birth: Evidence and Debate.* 2nd ed. London: Elsevier, pp. 97–113.

Csikszentmihalyi, M. 1975. *Beyond Boredom and Anxiety.* San Franciso, CA: Jossey Bass.

Davis-Floyd, R. 1994. 'The Technocratic Body: American Childbirth as Cultural Expression', *Social Science and Medicine* 38(8): 1125–40.

————— 2003. *Birth as an American Rite of Passage,* 2nd edn. Berkeley: University of California Press.

Downe, S. 2004. 'The Concept of Normality in the Maternity Services: Application and Consequences' in L. Frith (ed.) *Ethics and Midwifery: Issues in Contemporary Practice,* 2nd edn. Oxford: Butterworth Heinemann, pp. 91–109.

Downe, S. et al. 2008. 'The Early Pushing Urge: Practice and Discourse', in S. Downe (ed.) *Normal Birth: Evidence and Debate.* 2nd ed. Oxford: Elsevier, pp. 129–48.

Duden, B. 1993. *Disembodying Women: Perspectives on Pregnancy and the Unborn.* Cambridge, MA: Harvard University Press.

Dykes, F. 2002. 'Western Marketing and Medicine: Construction of an Insufficient Milk Syndrome', *Health Care for Women International* 23(5): 492–502.

————— 2005. 'Supply and Demand: Breastfeeding as Labour', *Social Science and Medicine* 60(10): 2283–93.

————— 2006. *Breastfeeding in Hospital: Midwives, Mothers and the Production Line.* London: Routledge.

Fairclough, N. 1992. *Discourse and Social Change.* Cambridge: Polity Press.

Flaherty, M. 1991. 'The Perception of Time and Situated Engrossment', *Social Psychology Quarterly* 54(1): 76–85.

Fox, M. 1989. 'Unreliable Allies: Subjective and Objective Time in Childbirth' in F.J. Forman and C. Sowton (eds), *Taking Our Time: Feminist Perspectives on Temporality*. Oxford: Pergamon, pp. 123–34.

Franklin, S. 1991. 'Fetal Fascinations: New Dimensions to the Medical-Scientific Construction of Fetal Personhood', in S. Franklin, C. Lury and J. Stacey (eds), *Off-Centre: Feminism and Cultural Studies*. London: Routledge, pp. 190–205.

Friedman, E.A. 1955. 'Primigravid Labor: A Graphicostatistical Analysis' *Obstetrics and Gynecology* 6(6): 567–89.

———— 1978. *Labour: Clinical Evaluation and Management*, 2nd edn. New York: Appleton-Century-Crofts.

Friedman, W. 1990. *About Time*. Cambridge, MA: MIT Press.

Gamble, J. et al. 2007. 'A Critique of the Literature on Women's Request for Cesarean Section', *Birth*. 34(4): 331–40.

Gaskin, I.M. 1990. *Spiritual Midwifery*, 3rd edn. Summertown, TN: The Book Publishing Company.

Goer, H. 1995. *Obstetric Myths Versus Research Realities: A Guide to the Medical Literature*. London: Bergin and Garvey.

Grimes, R.L. 2000. *Deeply into the Bone: Re-inventing Rites of Passage*. Berkeley, CA: University of California Press.

Haraway, D.J. 1991. *Simians, Cyborgs and Women: The Reinvention of Nature*. New York: Routledge.

Hashimoto, N. 2006. 'Women's Experience of Breastfeeding in the Current Japanese Social Context: Learning from Women and their Babies'. Ph.D. dissertation. London: Thames Valley University.

Helman, C. 2000. *Culture, Health and Illness*, 3rd edn. Oxford: Butterworth-Heinemann.

Jordan, B. 1997. 'Authoritative Knowledge and its Construction', in R.E. Davis-Floyd and C.F. Sargent (eds), *Childbirth and Authoritative Knowledge*. Berkeley: University of California Press, pp. 55–79.

Kirkham, M. 1983. 'Labouring in the Dark: Limitations on the Giving of Information to Enable Patients to Orientate Themselves to the Likely Events and Timescale of Labour', in J. Wilson-Barnett (ed.) *Nursing Research: Ten Studies in Patient care*. Chichester: John Wiley, pp. 81–99.

Lavender, T. 2003. 'NCT Evidence Based Briefing: Use of the Partogram in Labour', *New Generation* 24 : 14–16.

Llewellyn-Jones, D. 1982. *Fundamentals of Obstetrics and Gynaeology, Volume 1: Obstetrics*, 3rd edn. London: Faber.

McCourt, C. 2006. 'Supporting Choice and Control? Communication and Interaction between Midwives and Women at the Antenatal Booking Visit', *Social Science and Medicine*. 62(6): 1307–18.

Martin, E. 1987. *The Woman in the Body: A Cultural Analysis of Reproduction*. Milton Keynes: Open University Press.

Mittendorf, R. et al. 1990. 'The Length of Uncomplicated Human Gestation', *Obstetrics and Gynecology* 75(6): 929–32.

Murphy-Lawless, J. 1998. *Reading Birth and Death: A History of Obstetric Thinking*. Bloomington: Indiana University Press.

Myles, M.F. 1958. *A Textbook for Midwives*, 3rd edn. Edinburgh: Livingstone.

———— 1975. *A Textbook for Midwives*, 8th edn. Edinburgh: Churchill Livingstone.

Nguyen, T.H. et al. 1999. 'Evaluation of Ultrasound Estimated Date of Delivery in 17,450 Spontaneous Singleton Births: Do We Need to Modify Naegele's Rule?' *Ultrasound in Obstetrics and Gynecology* 14(1): 23–8.

Oakley, A. 1986. *The Captured Womb: A History of the Medical Care of Pregnant Women.* Oxford: Blackwell.

O'Driscoll, K., D. Meagher and P. Boylan. 1993. *Active Management of Labour*, 3rd edn. London: Mosby.

Petchesky, R. 1987. 'Fetal Images: The Power of Visual Culture in the Politics of Reproduction', *Feminist Studies* 13(2): 263–92.

Philpott, R.H. 1972. 'Graphic Records in Labour', *British Medical Journal* 21(833): 163–65.

Philpott, R.H. and W.M. Castle. 1972a. 'Cervicographs in the Management of Labour in Primigravidae, I: The Alert Line for Detecting Abnormal Labour', *Journal of Obstetrics and Gynaecology of the British Commonwealth* 79(7): 592–98.

———— 1972b. 'Cervicographs in the Management of Labour in Primigravidae, II: The Action Line and Treatment of Abnormal Labour', *Journal of Obstetrics and Gynaecology of the British Commonwealth* 79(7): 599–602.

Rothman, B.K. 1994. *The Tentative Pregnancy*, 2nd edn. London: Pandora.

Saunders, N. and C. Paterson. 1991. 'Can We Abandon Naegele's Rule?' *Lancet* 9(8741): 600–601.

Schwarz, E. W. 1990. 'The Engineering of Childbirth: A New Obstetric Programme as Reflected in British Obstetric Textbooks, 1960–1980', in J. Garcia, R. Kilpatrick and M. Richards (eds), *The Politics of Maternity Care*. Oxford: Oxford University Press, pp. 47–60.

Simonds, W. 2002. 'Watching the Clock: Keeping Time During Pregnancy, Birth, and Postpartum Experiences', *Social Science and Medicine* 55(4): 559–70.

Stacey, J. 1997. *Teratologies: A Cultural Study of Cancer.* London: Routledge.

Studd, J.W. and R.H. Philpott. 1972. 'Partograms and Action Line of Cervical Dilatation', *Proceedings of the Royal Society of Medicine* 65(8): 700–1.

Sweeney, K. and F. Griffiths. 2002. *Complexity and Healthcare: An Introduction.* Oxford: Radcliffe.

Tew, M. 1995. *Safer Childbirth? A Critical History of Maternity Care.* London: Chapman and Hall.

Thomas, H. 1992. 'Time and the Cervix', in R. Frankenberg (ed.) *Time, Health and Medicine*. London: Sage, pp. 56–67.

Wagner, M. 1994. *Pursuing the Birth Machine.* New South Wales: ACE Graphics.

World Health Organization. 1994. 'World Health Organization Partograph in Management of Labour', *Lancet* 343: 1399–404.

## CHAPTER 4

# THE PROGRESS OF LABOUR: ORDERLY CHAOS?

*Clare Winter and Margie Duff*

## Introduction

This chapter continues the theme of 'time' in childbirth developed in Chapter 3 with a focus on labour and birth. It looks at the findings from two studies of midwives' practices around assessment of progress in labour: one, a study of independent midwives' practices in England (Winter 2002); and second, an Australian study where the authoritative model of childbirth is medical and the accepted place of birth is in hospital (Duff 2005). Both studies show how midwives working to an alternative model, which refers to traditions of midwifery but in a postmodern context, conceptualize and respond to the labour process, resulting in a very different approach to time in childbirth.

The chapter begins with a brief review of midwifery practice before providing an overview of the two studies on which the discussion draws. It then describes the various authorized methods used to assess the progress of labour in modern and postmodern settings. It considers how 'countable time' has emerged as the guide to control labour within institutional birth settings and how health professionals working outside these institutions allow women 'uncountable time' and 'uncountable space' to labour and birth. Current methods used to assess the progress of labour include: monitoring the strength and length of contractions; observing the descent of the presenting part of the baby in relationship to the maternal pelvis; and digital internal vaginal examinations to assess cervical dilation and descent of the presenting part. The studies explain and provide examples of how midwives use clues to progress that encompass the whole woman.

That is, they assess each woman individually by observing the woman's response to her labour.

## Midwifery Practice

Midwives are health professionals that fulfil certain criteria outlined by the International Confederation of Midwives (ICM). These include defining the type of education, role and scope of practice a midwife may undertake and ensuring that, once it recognizes the person as successfully completing the educational requirements, each country places the person on a professional register.[1] Once registered, midwives are then able to work in a variety of settings, either as independent or as employed practitioners. For example, in the United Kingdom midwives may work in paid government positions in the community or in hospitals. Independent midwives provide continuity of maternity care to childbearing women (Winter 2002). Independent midwives are, therefore, autonomous practitioners who care for women throughout the childbirth experience.

In New Zealand, legislation in 1990 ensured that midwives could work autonomously within the healthcare system. These midwives provide continuity of care to women throughout the childbirth experience, including home or hospital care, and are known as Lead Maternity Carers (LMCs). However, midwives also work as core midwives, who staff secondary and tertiary hospitals and some birthing units. They provide twenty-four hour care to women in these facilities, including care to women who may have a doctor as their LMC (Guilliland, Tracy and Thorogood 2006).

In Australia, a medically dominated healthcare system exists providing few choices for midwives to work autonomously, although this is changing (Guilliland, Tracy and Thorogood 2006). Therefore, in Australia the majority of midwives work shifts in maternity and obstetric hospitals where the responsibility for the woman's care rests with medical staff.

## The Two Studies

### Contexts

The two studies by Winter (2002) and Duff (2005) explore the progress and assessment of labour. Winter's study examined independent midwives' practices, particularly those with respect to concepts and management of time. It focuses on some of the ways they assess the progress of labour. For these midwives, it is not important how long the labour takes, which is such a feature of hospital birth, but rather what matters is the process. Labour starts but as to how this journey proceeds is an unknown and uncertain passage and finishes with the birth of the baby. Uncertainty is a constant companion on this journey and midwives have become comfortable with this ambiguity.[2] The woman and the midwife embark on a journey that starts at booking and is completed when the woman is discharged from the midwife's care. The relationship that develops during the early part is one of the factors that lies at the heart of the trust that builds up through the journey. This partnership and trust allows the woman to move 'outside time' during labour. Assessing progress is a shared responsibility between the woman and the midwife (Winter 2002; Winter and Cameron 2006).

Duff's study of assessing cues to the progress of labour was set in Australian hospitals where midwives are required to work within protocols and polices around the medical assessment of labour. These hospital midwives therefore worked 'within time' while the responsibility of managing the progress or otherwise of labour rested with medical staff.

*Approach and Methods*

Winter's (2002) study was carried out for a Master's dissertation. At that time, very little research had been conducted on how midwives assess the progress of labour and no work had looked at how independent midwives assess the progress of labour. So the study was exploratory in nature and a qualitative approach was chosen since this tends to be holistic, striving for an understanding of individuals. It has an inherently flexible design that allows the researcher to systematically adjust during data collection as the emerging information brings up more questions (Strauss and Corbin 1998). The method was based on some of the principles of grounded theory, which provides a method to understand behaviour in different ways and reveal a pattern or model that the midwives involved could identify with. A purposeful sample of independent midwives was chosen since the individuals selected contributed a wide spectrum of experience and opinions to the study. Although only six midwives were interviewed, this represented eighteen percent of all known independent midwives in Britain at that time, a figure which has now trebled.[3]

Unstructured interviews were used so as not to influence the range and depth of the interviewees' responses, though with the use of grounded theory, as categories are identified, the interviewees were guided and probed more deeply during the interview in order to saturate categories. Some open-ended questions were designed to act as a means of triggering a response, allowing the interviewer to focus on particular issues if they did not come up spontaneously during the interview (Rose 1994).

Duff's (2005) doctoral study was developed to explain why some midwives appeared to explain the progress of labour in colloquial terms rather than the biomedical terms used within hospital settings. She often heard midwives say 'We won't be long' when they wanted to let another midwife know that they would soon need help at a birth. Sometimes, midwives got it wrong and birth would take much longer than anticipated but it appeared that it always seemed that the same midwives 'knew' while others did not. Why then did some midwives intuitively 'know' when a woman would give birth? When she casually asked these midwives how they 'knew' they just said they did. Others stated the women 'looked' like they would not be long (Duff 2005: xvi).

This led Duff to explore the literature on changes in women's behaviour during labour. She found that many texts by American nurse-midwives during the 1970s and 1980s used descriptions of women's responses to labour to assess labour progress. These descriptions were set within a framework of stages of labour developed by Friedman (1955), discussed in Chapter 3. A content analysis of these texts provided a rich source of data, which formed the basis for the development of a specially constructed labour assessment tool (LAT). The LAT was a combination of labour observations currently recorded on the partograph and descriptor cues from the content analysis.[4] The LAT was designed so that an observer could mark off the appropriate labour observations during set periods and was refined by an expert group process. After much discussion a decision was made that the midwives caring for the woman, and not an independent observer, would collect the data. This resolved the ethical dilemma of having a person unfamiliar to the labouring woman and her family present during the private birthing period. It also meant that the study could be larger than originally planned. The study was conducted in two Australian hospitals between 1999 and 2002, and collected the behavioural responses to labour of 179 consenting participants (94 primiparous and 85 multiparous women).

### Broad Findings

Winter (2002) identified many of the skills independent midwives use. She found that independent midwives have turned their back on the protocols of the medical model preferring to use midwifery skills to assess the process of labour. These skills include not only the interpretation of visual signs that change during labour but also decision-making skills based on midwifery knowledge and intuitive knowledge.[5] The midwife's ability to judge the progress of labour is greatly enhanced by the type of care the woman receives. The relationship between the midwife, woman and partner form a very important part of the process. Winter (2002) developed a model demonstrating how the midwives weave various aspects of knowledge together: moment by moment changes are processed internally and continue to satisfy the midwife that the labour is progressing and that the uncertain journey they are on is still safe. The precise length and timing of labour are not important factors in determining its progress as the focus is more about completing a process instead of achieving time limits on a chart. The midwives rarely use vaginal examinations since this will only inform them that the cervix is dilating. Rather, their skills are brought to bear on how the woman as a whole is progressing through labour, physically, mentally and spiritually. They are concerned that the whole person, rather than just the cervix, encounters labour (Winter 2002).

Duff (2005: 241), meanwhile, found that attempting to categorize behavioural responses into the phases of labour described by Friedman

(1955) provided few useful cues to assessing labour progress. When the results were viewed across time in labour, it was clear that maternal responses changed with 'time' in labour and that the responses differed with parity and whether the women's labours were induced or not. Duff argued that these changes were audible and visual physiological cues. These responses or cues changed during what Duff termed the 'starting out' period; the 'getting into it' period; the 'getting on with it' period; when women are 'nearly there' and again when 'the end is in sight' (2005: 240). Duff proposed that these terms, used in everyday midwifery language, describe the process and progress of labour more realistically than the medically defined stages of early labour, active labour, transition and second stage, which are based on divisions of cervical dilation measurement. She developed a model of labour that was not linear; that expanded or contracted and was based on a woman's physiological response to her labour (2005: 253).

The findings from these two studies involved skills of listening and observing. Both studies reported that the progress of labour is a complex, chaotic physiological process and not just based on 'time'. Both studies developed models to demonstrate the results from the data analysed.

## Comparing the Findings with Current Medical Methods

Current methods used to assess the progress of labour, in hospital settings worldwide, are briefly reviewed in this section and interspaced with findings from the studies of Winter (2002) and Duff (2005). All the following assessment methods, except for observing and monitoring the purple line and assessing descent through tracking changes in the unborn babies heart sound positions, are recorded on the partograph (see Chapter 3, this volume). These two assessments are described shortly.

In order to identify if a labour is progressing 'abnormally' in time women need a 'starting' point. This can be problematic for women birthing in hospitals. For example, in a study of the introduction of a new round partograph design (described below), Wacker, Kyelem and Bastert (1998) identified that 70 per cent of women stated they started labour 'today'; 22 per cent 'yesterday'; 4 per cent 'the day before yesterday'; and 2 per cent 'one month ago'. The last three responses do not fit with the 'scientific' concept that labour is a continuous (and short-term) linear process. Therefore women's responses to the question 'when did labour commence?' are either confirmed or discounted by professional staff (Enkin et al. 2000). Since health professionals find it difficult to accept women's statements of when labour started they have adopted the time women admit themselves to hospital as the 'start' of labour.

Partographs were developed as a rectangular graphical tool to chart the progress of labour based on Friedman's (1955) work and refined by a number of authors (see, for example, Glick and Trussell 1970; Studd 1973; WHO 1994). These refinements included an 'alert' and 'action' line. The 'alert' line is designed to signal to health professionals that the woman's labour may not be progressing swiftly enough and is a signal to transfer the woman for immediate medical assistance while the 'action' line indicates the need for prompt delivery of the baby. Working in West Africa, Wacker, Kyelem and Bastert (1998) argued that the rectangular partograph was too complex for staff to use and that transfers to a major hospital were not instigated when the alert line was crossed. Furthermore, they found that 97 per cent of women whose labours crossed this point birthed normally.

These issues prompted Wacker, Kyelem and Bastert (1998) to develop a simplified round partograph based on a twelve-hour clock pattern (see figure 4.1). This round partograph consisted of two parts. The first is a rectangular sheet of paper with the usual hospital labour and birth information (for example name, age, date and time of arrival, number of pregnancies, when labour began) and in the centre a large diagram of a clock. The clock is segmented into twelve sections with each segment representing one hour and labelled like traditional clocks. Within the clock are ten decreasing circles representing cervical dilation, each numbered

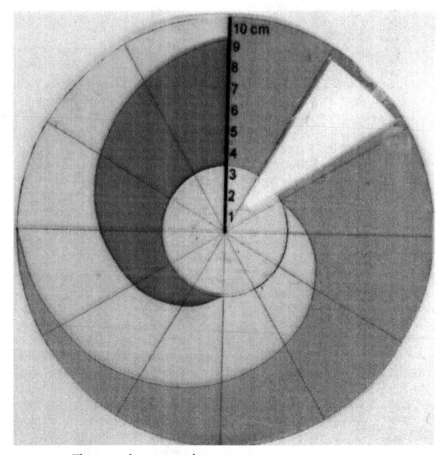

Figure 4.1: The round partograph.

accordingly from one (at the 'bulls eye' position in the centre of the clock) to 10 centimetres just below the 12 o'clock point. On arrival at hospital the woman has a vaginal examination to determine her cervical dilation. The result (in centimetres) is marked on the clock in relationship to the time (for example, 3 cm at 2 o'clock). The results of each subsequent vaginal examination are also recorded. The second part consists of a separate transparent dial, the same size as the 'clock', with a darker straight line at the 12 o'clock position and the same 1–10 cm segments representing cervical dilation. It has three coloured coded spirals (green, yellow and red), respectively representing 'normal' progress, an 'alert' stage and need for 'action'. The green spiral, for example, begins at 3 cm in what would be the 12 o'clock point with its inside edge at the junction of 3 cm and at the 2 o'clock point. It swirls out to meet the 10 cm circle at the 9 o'clock point. Each time a vaginal examination is made and recorded on the paper clock, the transparent dial is placed over the paper clock with its black line laid over the admission time so that the result is seen clearly in either the green, yellow or red zone.

Although the round partograph continues the concept of limiting women's labours through clock time, it takes a less linear form and this appeared to function more effectively from the professionals' viewpoint. It also allows more clock time for labour, prolonging the time limits established by the WHO partograph for early labour (latent phase) from eight to twelve hours. Furthermore, it delays the alert and action lines by a further two hours each. Wacker, Kyelem and Bastert's (1998) study compared the round partograph with the rectangular WHO partograph. Both partographs were completed on each woman giving birth over a period of three months. The authors analysed staff satisfaction and birth outcomes from the partographs of 203 women. They found 86 per cent of staff preferred the round partograph and found it easier and quicker to use. Furthermore, they found that 22 per cent of alert lines were crossed in the round partographs compared to 78 per cent in the rectangular partograph yet this did not alter birth outcomes. Compared to the previous non study period there were no statistically significant differences in the number of neonatal and maternal deaths or the number of caesarean sections. Therefore, allowing women more 'time' to birth appears to make no difference to these life outcomes even though it was designed for use in settings where remoteness may mean slow and difficult access to a hospital if problems arise.

## Monitoring Contractions

Painful contractions signal to women that labour and birth are approaching and that they should begin their preparations for this event. The actual time at which labour culminates, however, is not always straightforward. In 1671, the English midwife Jane Sharp mused on the fact that 'it is hard to know when the true time of her travel is near, because many women have great pains many weeks before the time of delivery comes' (Sharp 1999[1671]: 159), but she provided cues to the commencement of labour: ruptured membranes, the shape of the abdomen (when lightening has occurred), and the type of contractions that arise in the top of the abdomen and extend down to the groin.

Throughout the nineteenth and twentieth centuries, uterine contractions were generally considered by medicine as 'abnormal' if they were not regular or consistently increased in strength as labour progressed (see, e.g., Calkins, Irvine and Horsley 1930; Friedman 1955; Llewellyn-Jones 1971; Playfair 1890). The midwifery literature also proposes that as labour progresses contractions become more frequent (see, e.g., Olds et al. 2004; Varney, Kriebs and Gegor 2004; Fraser and Cooper 2003; Nicols and Zwelling 1997; Gorrie, McKinney and Murray 1994). Few authors, however, identified differences in contractions between non-induced and induced labours. For example, McKay (1994) did note that inductions of labour cause contractions to be closer together in early labour even though there is little cervical dilation occurring (see also Chapter 9, this volume).

Winter (2002) reports that independent midwives do not consider contractions to be the most important way of assessing the progress of labour. Nonetheless, contractions are an integral part of and essential to labour. All the midwives in her study considered the 'rate, strength and frequency of contractions' as 'standard stuff', but the factors which indicate actual progress are not easily definable. So, as long as there are contractions there will be progress and time will tell if there is no progress. The effectiveness of contractions is regarded as important but there are no clear guidelines as to what constitutes an effective contraction. As long as contractions are regular and the woman is able to concentrate and cope with them, 'that would count', as indicated by midwives in her study:

> Concentrating on what they are doing and the contractions are coming regularly, they don't have to be every two minutes or every five minutes or every six minutes, normally they are less than every ten minutes, but you could have somebody getting contraction every ten minutes that were really, really intense and were lasting a long time and that would count, I would be thinking, OK she probably is in good labour. (Winter 2002: 60)

The woman is coping really well with strong contractions that are obviously very powerful, that she is really into the rhythm of it, she is doing well, then I think I'm very comfortable with that. (Winter 2002: 60)

These comments are supported by Duff's (2005) study, which found that women who commenced labour spontaneously had less frequent contractions than women who were induced into labour. Duff reported that 35 per cent of the non-induced primiparous women's contractions in her study became spasmodic and less frequent at varying times throughout labour although the women all went on to labour and birth without interventions. Aderhold and Roberts (1991) also reported that contractions can become infrequent during the first stage of labour. When this occurs however, it is considered 'abnormal' by other authors and labelled 'dysfunctional' (Varney, Kriebs and Gegor 2004), 'protracted' (Enkin et al. 2000), or as 'inertia' (Friedman 1955) and 'lingering' (Playfair 1890). As Enkin et al. note, this type of labour 'has been recognized as a problem for centuries' (2000: 334). Rather than being necessarily problematic, this may simply be a physiological function. To make a judgement that it is problematic requires additional evidence. For example, if the contraction pattern is causing maternal or physical fetal distress there should be clinical signs. If the problem was dehydration this would be observed through decreased fluid intake during the early part of labour, vomiting, ketonuria, an increased temperature and an increased pulse rate.

The midwives interviewed by Winter (2002) did not only consider physical reasons why contractions were slowing down or disrupted. They believed this situation could be caused by the woman's emotional or psychological state of mind. They also related how a change in the environment, such as a disagreement between the woman and her partner or loss of trust between the woman and the midwife, could also affect contractions and disrupt labour. There is physiological evidence to support the theory that stress and anxiety will interfere with the process of labour via the secretion of oxytocin (Zac, Kurzban and Matzner 2004):

> If somebody is struggling to cope with contractions, um, even after you  know, they have been going for a while, and they are struggling to get into the rhythm of it for whatever reason, I might then, part of me, I will have antenna I think that will say okay, what, why is she not able to. And it may be that, um, you know there's [something] going on in her head about, maybe its fear, maybe … if its her first baby, is this about, you know, becoming a mother is there an anxiety. (Winter 2002: 61)

The midwives felt they were able to detect the subtle changes in the intensity of contractions that indicate progress because they stay with the woman. The midwives make sense of these changes and they are able to put them in the whole context of the labour. The ability to interpret the changes taking place in the labour is attributed to having cared for the woman throughout her pregnancy. Having a trusting relationship with the woman is seen as one of the pivotal reasons for how their midwifery skills work:

> As contractions go off or she's finding things really hard work but that doesn't seem to be progressing, you know, you expect her to be getting up to fully but that doesn't seem to be happening or something like that maybe … I think because of the way we work we really know who we're working with that makes a big difference as well, so that makes things a lot easier and I think you have a lot much more kind (sic) open communication so very often I'll ask woman what they're feeling, and that gives you good clues as to what is actually happening or they'll tell you what is happening if you give them an opportunity. (Winter 2002: 62)

Albers (2001) identified that protracted or prolonged labour, failure to progress, slow labour and dysfunctional labour make up the category known as 'dystocia' and this diagnosis is the most frequent reason for augmentation of labour with the synthetic hormone oxytocin and for caesarean sections. Yet these longer 'on and off again' labours may be more gentle and are not necessarily associated with poor outcomes Albers (2001). The evidence from the studies by Duff (2005) and Winter (2002) highlight that contractions can become less frequent, become less intense or stop at varying times during labour. These 'lulls' may be 'physiological' and non-problematic in occurrence without other signs or symptoms of problems.

Midwives identified a noticeable change around what is conventionally known as 'second stage'. This phase seems to herald a change in behaviour as it is a period where the cervix is finally pushed back by the advancing fetal head (Winter 2002). Often at this point the woman feels she can no longer cope, and this is associated with changes in her behaviour. Again, this information is interpreted in the context of the whole process of labour, contractions are seen in context with other signs to indicate the approach of second stage:

> Often they're saying I can't cope, they've had enough and all that kind of stuff, and when the woman has been coping quite well it often makes you think well you know, you're probably heading towards the end of first stage, um, and putting all that in context with what's happening with the contractions. (Winter 2002: 60)

Although no specific statements were recorded as being made by the women at any particular time period during labour in Duff's (2005) study, there were behavioural changes associated with speech which act as cues to progress. For example, during the 'getting on with it' stage the majority of women were recorded as having a 'sharp and snappy' tone of voice whereas they were recorded as having a 'quiet subdued' tone previously in the 'getting into it' stage. This changed to a 'demanding' or 'loud voice' for the majority of multiparous women during the 'nearly there' stage whereas the primiparous women continued to have a 'sharp, snappy' tone of voice (Duff 2005: 245).

### Assessing Fetal Descent Abdominally in Relationship to the Maternal Pelvis

There are two methods that may be used to abdominally assess descent of the fetus during labour. The first is to track changes in the position of the fetus in relationship to the maternal pelvis through palpation, and the second is to track the corresponding changes in movement of the fetal heart sounds (Munro Kerr et al. 1939).

Assessing the progress of labour by determining descent of the 'presenting part' during vaginal examination was examined extensively by Friedman and Sachtelben (1965a, 1965b). Their study of over 1,000 women's labours suggested that a high presenting part (measured as centimetres above the ischial spines of the pelvis) prolonged the latent phase of labour and resulted in a higher incidence of dysfunctional labour. No detail was provided as to how the data were collected and from whom, or under what conditions, apart from the use of frequent 'cervical assessments' (vaginal examinations). Winter (2002) found that midwives made use of the fact that the fetal heartbeat can be traced across the surface of the abdomen as the baby descends during labour and 'tucks behind the pubic bone' in late labour.[6] If the distinctive sound of the umbilical cord can be picked out in the pubic area, this informs the midwife that the cord is round the neck and prepares her for that. When the fetal heart rate drops towards the end of labour, it is interpreted as the head descending in second stage:

> If you've got it and you've had it's clear, and it's still clear, and you know, it will move down as the baby's moving down, so that, that can be really helpful, as can picking up the cordal sounds around the pubic hair and thinking I'll be prepared for the baby to come out with the cord round the neck and those kind of things. (Winter 2002: 58)

Assessing descent abdominally through palpation was also a feature of Duff's (2005) study. However, this was one aspect that was not recorded by the midwives on a regular basis after the initial assessment, suggesting they did not consider it useful during normal birth. Moreover, from clinical experience, they reported that women complain that this procedure is painful during labour. Descent of the presenting part is also dependent on the type of pelvis. Studies have shown that the non-engagement[7] of the fetal head at the onset of labour is not uncommon (Deitze 2001). Glick and Trussell (1970) noted that, in Uganda, primigravid women rarely had engaged presenting parts at the onset of labour as women generally had shallow pelves. Friedman and Sachtleben's (1965a, 1965b) study resulted in clinicians considering every labour to be 'a trial of labour', even though it did not take into account the cultural basis of the research on which their views of normal labour were based. The clinical application of these results was the inclusion of this method of assessing descent of the fetal presenting part into the partograph together with estimating cervical dilation, using norms that were assumed to be universal.

### *Observing and Monitoring the Purple Line and the Rhombus of Michaelis*

A possible method for assessing the progress of labour without resorting to invasive and frequent cervical examinations is to observe the purple line (Byrne and Edmonds 1990) and the Rhombus of Michaelis (Sutton and Scott 1995). Hobbs (1998) described the purple line phenomenon and how it appears at the edge of the anal margin when the cervix is about two centimetres dilated. The line then follows the natal cleft as cervical dilation increases and ends at the nape of the buttocks when the woman is fully dilated. Most midwives in Winter's (2002) study spoke of the purple line that extends gradually from the anus to the sacrum as labour progresses. Generally they considered it to be unreliable except if the woman is in second stage. Since many of the women they cared for in labour tended to be on all fours, it is quite easy for this sign to be observed without the woman feeling they are being deliberately watched.

There is also an area above the ligaments connecting the sacrum to the ilia that becomes more flexible during labour, known as the Rhombus of Michaelis (Sutton and Scott 1995). When a woman is on her hands and knees or standing, the Rhombus of Michaelis can be seen clearly because of pressure from the fetal head. This feature is apparent just prior to the beginning of the involuntary pushing urge of the second stage of labour when the fetal head lifts the sacrum and coccyx out of the way. The women in the independent midwives' care rarely give birth lying on a bed so the midwives have a good view of the woman's back. If women are clothed, they generally strip off around this time:

> I try to use the purple line, I have varying success with it ... I think it's one of those things that, you really can see it when somebody is fully, fully dilated, sometimes in the process of getting there it can be misleading. (Winter 2002: 58)

> Then there's the good old red line that goes up the back but you don't see that until fairly late on in labour which is when women start to strip off and get on all fours or they are wanting to get in the pool or something like that you know. (Winter 2002: 58)

Personal experience of one author (Duff 2005) suggests that this line is not always visible on women with darker skin. The different methods of assessment used by the midwives studied, in contrast to those in mainstream obstetrics, illuminate the tendency of medical assessments to focus on certain senses or concepts and forms of measurement rather than others.

### Vaginal Examinations

Vaginal examinations are used to determine cervical dilation, confirm fetal presentation and position, and assess progress in labour and estimate descent in relation to other internal pelvic landmarks (Fraser and Cooper 2003). The independent midwives in Winter's (2002) study do not use vaginal examinations (VEs) to assess the progress of labour. The main function of vaginal examinations in their practice is to clarify a situation or confirm what is already suspected, contrary to hospital policies where vaginal examinations are almost exclusively used to assess progress. Vaginal examinations were sometimes performed if the midwife was considering leaving the woman for a while and, used along with other signs, helped the midwife make an assessment of whether she should leave or not. The results of vaginal examinations are considered in conjunction with the other signs and what the woman is saying. The midwives regard the results of a vaginal examination as having equal status with the other signs and what the woman says. They find vaginal examinations particularly useful when there is a conflict of information, as when signs suggest one thing whilst the woman herself claims the opposite. Once again, examinations are used as a tool to clarify rather than to predict or measure progress:

Well I'd say I probably get through eighty to ninety percent of labours without a VE, so that sounds like not very useful but ... I would not like to go to a labour without that in my tool bag because I think it's when you need it. (Winter 2002: 63)

... when you have a conflict between the information you are receiving, if somebody tells you that they are labouring well and hard and you don't think they are, or somebody tells you they are wanting to push and you don't think they are, or, somebody says they are not pushing and they are, you know, it's that, they are showing you all the signs of doing something, but they say they are not, then the VE's useful. ... a bit like ultrasound scans when there's a discrepancy between what you feel [abdominally] and the information you have from last menstrual period or something like that ... you say okay let's use a tool to see which one it comes up with, and probably the VE is like that. (Winter 2002: 63)

Although VEs have their place in assessing labour, routine examination is viewed as disruptive to labour and an unnecessary intrusion used to complete a chart:

To get a good partogram reading you need regular VEs and I think they are intrusive and they may stop people labouring. (Winter 2002: 65)

There is also a strong feeling about conducting needless vaginal examination on a woman. One midwife made her position clear: she asks herself whether an examination is really necessary or if it can be avoided, as this will enhance the woman's sense of achievement because she has trusted her own body. Inappropriate vaginal examinations are seen as potentially abusive, and this has consequences for the woman experiencing the procedure and so they should not be undertaken lightly. The distress caused to women by the invasive nature of vaginal examinations has been pointed out by a number of writers (Bergstrom el al. 1992; Clement 1994; Devane 1996; Warren 1999).

The midwives on the whole saw vaginal examinations as a fallback when all else fails rather than a first resort, particularly to be used if the midwife is tired and her other senses are failing, when she becomes more ritualistic in her care. It is useful in an emergency, though 'emergency' was not defined. They also identified vaginal examinations as typifing the medical model; they are relied on when the midwife does not have a trusting relationship with the woman and her partner. This would suggest that not knowing the woman must in some way blunt their awareness of the other signs used to assess the progress of labour – not knowing the woman engenders an environment of uncertainty or it might alternatively reflect the privileging of certain forms of observation, senses or knowledge over others, in biomedicine and in modern society. The midwives also associated the need for regular vaginal examinations as being a defensive practice: within our litigious society, the NHS deems that the recording of VEs indicates good care for the woman. It is also seem as an objective measure (distance against time) recorded in the medical notes and provides clinicians with justification for intervention if adequate progress has not been achieved.

The aim of Duff's (2005) study was to examine behaviours exhibited by women during labour together with routine partograph observations. Hospital partograph polices influenced the study, with vaginal examinations being undertaken, with the woman's consent, on admission to hospital, then every four hours and again to confirm full dilation. Although it was not the focus of the study to detail 'mean' labour curves it is worth reporting that there were patterns found in dilation. Altogether, 18 per cent of primiparous women and 8 per cent of multiparous women had cervical dilation measurements which did not change on the next examination. That is, a plateau occurred during the dilation process. This challenges the idea that the progress of labour is linear. Interestingly, while six primiparous women had epidural blocks inserted after the second cervical examination, all the rest laboured and birthed without assistance. Why these plateaux occurred is not known. Winter (2002: 126) argues that the examination itself may disrupt labour, though these plateaux may also be part of a normal process. Duff (2005: 285) suggests that when contractions ease dilation may slow and act as 'rest' periods for the woman and her baby during labour. The problem of comparing all women's labours to a 'mean' curve dismisses other plateaux that occur during individual women's labours. Friedman (1955) defined these plateaux during labour as 'clinical inertia', noting that many were caused by problems associated with excessive medication, disproportion (where the baby's head is larger than the woman's pelvis) and posterior positions (where the baby lies with its spine along the mother's spine, often causing pain because of pressure exerted on spinal nerves), although there were 20 per cent of his primigravida group in which no cause could be identified. It is further argued that plateaux create problems for predicting and controlling time in labour. Plateaux disrupt the linear model of labour and suggest there are other complex issues involved.

### Putting the Woman Back into Her Labour

The partograph and its recorded observations ignore the woman as a person responding to her environment and her body. The following examples from Winter's and Duff's studies provide clues to how midwives assess such responses. From the data collected by Winter (2002), central to the practice of care is the trust relationship between the woman and the midwife. Much of the ability to interpret labour is based on the value of the style of care independent midwives give. Another important factor in interpreting the signs of labour is the midwife's ability to be 'absorbed' in the process. Their presence is unobtrusive: they focus on becoming one with the process itself. In some respects, the midwives allow their senses to be immersed in what is happening:

> … there's a thing about being still and giving your time to listen to the labour, um, so listening to the labour, and listening to the woman, and I think I'm

only becoming newly aware of how much you pick up in your own body and about the body you're with, and that sounds a bit disconnected, but I think that ... midwives, particularly independent midwives ... is that you do you actually listen to bodies, those kind of things tell you quite a lot. (Winter 2002: 106)

... becoming part of the scene by absorbing, you're letting the information that you are receiving from your senses wash over you but you're also becoming part of the scene so you're becoming absorbed into it. ... You're not intruding on the situation, you're becoming part of it, which probably makes it easier for you to observe without the woman's looking like (sic) feeling like she's in a goldfish bowl being looked at, that's probably what I' meant by absorbing. (Winter 2002: 107)

Anderson (2000), herself an independent midwife, highlighted in her study of women's experience of labour that the midwife who facilitates an 'unobtrusive atmosphere' enables the woman to feel safe enough to abandon herself to the process, whereas the 'intrusive' midwife disables the woman, turning her birth into a 'nightmare'.

Uncertainty was another important element in the care provided by the midwives. Even though they trusted the woman and their own midwifery skills, they acknowledged that the process of labour is uncertain. Rather than the rigid partogram used in the medical model, the midwives were comfortable with the possibility that what they were 'sensing' may not be correct:

Noise gives very good clues as to what the woman were doing, they are very often [but] not always a sure fire certainty. (Winter 2002: 109)

The way she's breathing [you can tell] to some extent how far on she is in labour, now I'd say nine times out of ten that's very accurate but I'd say there are some women and you haven't got a clue. (Winter 2002: 110)

I've known what their voice sounds like and so I can pick up initially from that. It doesn't always work. (Winter 2002: 110)

I have learnt you can have all sorts of signs of imminent second stage and it's absolutely nowhere near. (Winter 2002: 64)

Breathing patterns were examined by Duff (2005). She found that patterns changed as labour progressed but there were differences as to when this occurred during labour between the parity groups. For example, the majority of nulliparous women were recorded (Duff 2005: 244) with deep and slow breathing patterns together with increased breathing patterns within the same contraction period during the 'getting into it' stage, while the majority of multiparous women were only recorded with deep and slow breathing patterns. These patterns changed during the 'nearly there' stage with the majority of both primiparous and multiparous women recorded as 'holds breath/pushing'. This descriptor emerged, in excess of 50 per cent of the time, an hour prior to full dilation in the primiparous group, but was recorded in 25 per cent of cases two hours before full dilation. In the multiparous women a similar pattern was seen but within a time frame of less than an hour prior to full dilation. Pushing prior to second stage has been well documented by a number of authors (see, for example, Davis 1992; Downe 2003; Olds et al. 2004; Varney, Kriebs and Gegor 2004).

Noise was also explored in Duff's study. She found that women 'moaned and groaned or made deep throaty noises' during the 'getting into to it' stage and this continued throughout labour (Duff 2005: 244). Two descriptors, 'grunting' in induced women and 'cries/yells' in non-induced women, were the only non-verbal sounds that emerged during the 'nearly there' stage but decreased in occurrence during the 'end is in sight' stage (Duff 2005: 245). Such audible signs may not be accessible or valid, however, in certain cultural settings. In some cultures – the United Arab Emirates, for example – it is disapproved of for birthing women to make noises during labour, while in the cultural setting of an obstetric hospital unit, where women often have epidural pain relief, audible expressions of labour are also seen as abnormal.

A further component of Winter's (2002) model was the need for the midwives to have a philosophy that labour and birth are a normal physiological process. This gives them confidence in letting the process happen rather than needing to control physiology:

> Normal labour for me is a labour that starts and a woman progresses at whatever rate. I'm not really bothered so long as she is progressing, and that the end of it she gives birth to live baby, I suppose that the progression is in line with kind of what's happening, with uterine activity really, so you know she's contracting one in five although through (sic) I don't expect to have a twelve-hour labour. (Winter 2002: 112)

> I would link it with what I feel is as a midwife and very much an instinctual midwife, I would [be] linking it with women's knowledge that goes back through the centuries. (Winter 2002: 113)

## Deliberations: Themes and Connections

The two studies under discussion have identified lineal 'time' as not being an important part of assessing the progress of labour, contrary to what authorized biomedical practice would suggest. Labour begins and labour ends but how and when that end is reached is complex and chaotic. It can be argued that in the context of childbirth both 'time' and 'space' are linked. For example, both time and space can be considered as countable (measurable) or uncountable. Countable time revolves around defining a specific point in time and measuring it against clock time while countable space involves measurable areas. In contrast, uncountable space can be identified as personal freedom, the space to be oneself. During labour this is the freedom to respond individually. Uncountable time is not measurable. For example, is it time for a rest? For pregnant women it can be seen as a time to birth. Only after time has passed can it be measured.

The major themes that emerge during the review of these studies include: labour and birth as a physiological process; the cues to some of these physiological changes; and the individuality of labours. Both authors argue that labour and birth is a physiological process but what marks the beginning of labour is problematic. For independent midwives identifying the 'start' of labour is not important as they do not measure clock time in labour. Although time must pass between the start and end of labour, the journey is not linear. The studies provide evidence of this. For example, Duff's (2005) study shows that cervical dilation plateaux occur in individual women's labours and she argues that, as with Friedman's findings, labour is not a linear process. Friedman's labour curve was a sigmoid shape, indicating that plateaux occurred at either end to creating the sigmoid shape. If there were no plateaux or changes in pace, the mean labour curve would be a straight line.

Furthermore, Duff (2005) provides evidence that contractions do not always increase in intensity and strength as labour progresses, as many texts contend. Winter's (2002) study considered other reasons why contractions slowed down or were disrupted. Independent midwives believed that the situation can be caused by a woman's emotional or psychological state of mind, or changes in a woman's environment, such as a disagreement with her partner or a loss of trust between the woman and the midwife. She argues that there is a physiological reason: stress and anxiety will hinder the secretion of oxytocin required to stimulate contractions (Winter 2002: 132). But it can also be argued that these form a complex association between events in uncountable time and uncountable space.

Birth is an important human process and takes uncountable and countable time: women need the 'space' in which to take this time. For example, Winter's (2002) study shows how independent midwives provide this space, allowing women to take their own time to birth. Midwives working in a hospital setting with policies that follow the World Health Organisation's policies on monitoring normal birth may have this ability restricted by policies and protocols. Likewise, women in Winter's study laboured and birthed at home in their own countable space (their home) and free to be oneself (uncountable space). In contrast Duff's study was situated in a hospital where space was restricted through the design of the birthing areas and by hospital policies. For example, women who are noisy in labour are offered epidurals to reduce their voices while hospital policies provide restrictions on women's uncountable time in labour through changing it to countable time.

Another reason why women are given epidurals during labour is to relieve them of pain. Often when women's labours are not progressing (that is there has been no increase in cervical dilation) medical staff may offer the woman an epidural and then stimulate her labour with an intravenous drug. In Duff's study only a few women who had unchanging cervical dilation curves had an intervention in the form of an epidural (Duff 2005: 187). The study did not identify what the midwives did to prevent or stop interventions happening. It could be argued that since these midwives had volunteered to collect the data they already had a philosophy of what was 'normal'. Hence, like midwives in Crabtree's (2008) study, the Australian midwives in Duff's (2005) research may have found other ways to work within a policy and maintain their philosophy. For example, one of the midwives who volunteered to collect the data confided to the author that she never started women on a partograph until they were 'pushing'. She only volunteered to collect the data as she believed that labour and birth were a normal process and that vaginal examinations were not necessary to assess labour progress.

In conclusion, it can be argued that Westernized medical settings provide a narrow and rigid set of parameters that have become the authorized approach to monitoring and assessing labour. Midwives, and in particular independent midwives, use a wider range of senses, and a more complex model of the processes involved in childbirth, with a less rigid and linear notion of time and space, which perhaps better reflects the complexity of the process of human childbirth.

## Notes

1. International Confederation of Midwives. 2006. 'Definition of a Midwife'. Retrieved 15 December 2006 from: http://www.internationalmidwives.org/.
2. Traditional midwives, practising worldwide and across history, were more comfortable in dealing with uncertainty, but it has been argued that biomedical and hospital-based practice reduces midwives' capacity to deal with the uncertainty of childbirth. Good accounts of this are given in Jordan (1993), for Yucatan and Holland as compared with the U.S., and in Olafsdottir (2006), for Iceland.
3. Independent Midwives Association. 2006. 'Register of Independent Midwives'. Independent Midwives Association (U.K.). Accessible via the Independent Midwives Association. Contact details can be obtained via: www.independentmidwives.org.uk
4. The cues included very detailed observation and recording of bodily responses, including posture, movement, sound and facial expression.
5. For a discussion of the concepts of intuition and expertise in midwifery, see Downe Simpson and Trafford (2007).
6. The midwives use a small stethoscope (called a Pinard) or a sonicaid to listen at regular intervals, rather than continual monitoring by abdominal belt from a cardiotacograph, as is common practice in maternity hospitals.
7. This is the term used to describe the point when the largest diameter of the fetal head has passed into the pelvis brim (Tiran 2003).

# References

Aderhold, K.J. and J.E. Roberts. 1991. 'Phases of Second Stage Labour'. *Journal of Nurse-Midwifery* 36(5): 267–75.

Albers L.L. 2001. 'Rethinking Dystocia: Patience Please', *Midwifery Information and Resource Service (MIDIRS) Midwifery Digest* 11(3): 351–53.

Anderson, T. 2000. 'Feeling Safe Enough to Let Go: The Relationship between a Woman and her Midwife During the Second Stage of Labour', in M. Kirkham (ed.) *The Midwife–Mother Relationship*. Basingstoke: Macmillan, pp. 92–118.

Bergstrom, L. et al. 1992. 'You'll Feel Me Touching You, Sweetie: Vaginal Examination During the Second Stage of Labour', *Birth* 19(1): 10–18.

Byrne, D. and D.K. Edmonds. 1990. 'Clinical Method for Evaluating Progress in First Stage of Labour', *Lancet* 355: 122.

Calkins, L.A., J.H. Irvine and G.W. Horsley. 1930. 'Variations in the Length of Labour', *American Journal of Obstetrics and Gynaecology* 19: 294–97.

Clement, S. 1994. 'Unwanted Vaginal Examinations'. *British Journal of Midwifery* 2(8): 368–70.

Crabtree, S. 2008. 'Midwives Constructing "Normal Birth"', in S. Downe (ed.) *Normal Childbirth: Evidence and Debate*. 2$^{nd}$ ed. Edinburgh: Churchill Livingstone, pp. 97–113.

Davis, E. 1992. *Heart and Hands*, 2nd edn. Berkeley, CA: Celestial Arts.

Deitze, M. 2001. 'A Re-evaluation of the Mechanism for Contemporary Midwifery Practice', *Midwifery Matters* 88: 3–8.

Devane, D. 1996. 'Sexuality and Midwifery', *British Journal of Midwifery* 4(8): 368–70.

Downe, S. 2003. 'Transition and the Second Stage of Labour', in D.M. Fraser and M.A. Cooper (eds), *Myles' Text for Midwives*, 14th edn. Edinburgh: Churchill Livingstone, pp. 487–506.

Downe, S., I. Simpson and K. Trafford. 2007. 'Expert Intrapartum Maternity Care: A Meta-synthesis', *Journal of Advanced Nursing* 57(2): 127–40.

Duff, M. 2005. 'A Study of Labour', Ph.D. dissertation. Sydney: University of Technology.

Enkin, M. et al. 2000. *A Guide to Effective Care in Pregnancy and Childbirth*, Oxford: Oxford University Press.

Fraser, D.M. and M.A. Cooper (eds). 2003. *Myles' Textbook for Midwives*, 14th edn. Edinburgh: Churchill Livingstone.

Friedman, E.A. 1955. 'Primigravid Labor: A Graphicostatistical Analysis', *American Journal of Obstetrics and Gynecology* 6(6): 567–89.

Friedman, E.A. and M.R. Sachtleben. 1965a. 'Station of the Fetal Presenting Part, I: Pattern of Descent', *American Journal of Obstetrics and Gynecology*, 93(4): 522–29.

———— 1965b. 'Station of the Fetal Presenting Part, II: Effect on the Course of Labor', *American Journal of Obstetrics and Gynecology*, 93(4): 530–36.

Glick, E. and R.R. Trussell. 1970. 'The Curve of Labour Used as a Teaching Device in Uganda', *Journal of Obstetrics and Gynaecology of the British Commonwealth*, 77: 1003–6.

Gorrie, T.M., E.S. Mckinney and S.S. Murray. 1994. *Foundations of Maternal Newborn Nursing*. Philadelphia, PA: Saunders.

Guilliland, K., S.K. Tracy and C. Thorogood. 2006. 'Australian and New Zealand Health and Maternity Services', in S. Pairman et al. (eds), *Midwifery: Preparation for Practice*. Sydney: Elsevier, pp. 3–33.

Hobbs, L. 1998. 'Assessing Cervical Dilatation Without VEs: Watching the Purple Line', *The Practising Midwife* 1(11): 34–35.

Jordan B. 1997. 'Authoritative Knowledge and Its Construction', in R.E. Davis-Floyd and C.F. Sargent (eds), *Childbirth and Authoritative Knowledge: Cross Cultural Perspectives*. Berkeley: University of California Press, pp. 55–79.

Llewellyn-Jones, D. 1971. *Fundamentals of Obstetrics and Gynaecology, Volume 1: Obstetrics*. London: Faber.

McKay, S. 1994. 'Labor: Overview', in B. Katz Rothman (ed.) *The Encyclopedia of Childbearing*. New York: Holt, pp. 212–15.

Munro Kerr, J.M. et al. 1939. *Combined Textbook of Obstetrics and Gynaecology*, 3rd edn. Edinburgh: Livingstone.

Nicols, F.H. and E. Zwelling 1997. *Maternal-Newborn Nursing: Theory and Practice*. Philadelphia: Saunders.

Ólafsdóttir, O. 2006. 'An Icelandic Midwifery Saga – Coming to Light: 'With Woman' and Connective Ways of Knowing'. Ph.D. dissertation. London: Thames Valley University.

Olds, S.B. et al. 2004. *Maternal-Newborn Nursing and Women's Health Care*, 7th edn. Upper Saddle River, NJ: Prentice Hall.

Playfair, W.S. 1890. 'A Discussion on Modern Methods of Management of Lingering Labour', *British Medical Journal* X (September 27th): 715–17.

Rose, K. 1994. 'Unstructured and Semi-Structured Interviewing', *Nursing Researcher* 1(3): 23–32.

Sharp, J. 1999[1671]. *The Midwives Book: Or, the Whole Art of Midwifery Discovered*. E. Hobby (ed.) Oxford: Oxford University Press.

Strauss, A. and J. Corbin. 1998. *Basics of Qualitative Research: Techniques and Procedures for Developing Grounded Theory*. London: Sage.

Studd, J. 1973. 'Partograms and Normograms of Cervical Dilation in Management of Primagravid Labour', *British Medical Journal* 4: 451–55.

Sutton, J. and P. Scott. 1995. *Understanding and Teaching Optimal Foetal Positioning*. Tauranga, New Zealand: Birth Concepts.

Tiran, D. 2003. *Baillière's Midwives' Dictionary*, 10th edn. Edinburgh: Baillière Tindall.

Varney, H., J.M. Kriebs and C.L. Gegor. 2004. *Varney's Midwifery*, 4th edn. Boston, MA: Jones and Bartlett.

Wacker, J., D. Kyelem and G. Bastert. 1998. 'Introduction of a Simplified Round Partograph in Rural Maternity Units: Seno Province, Burkina Faso, West Africa', *Tropical Doctor* 28: 146–52.

Warren, C. 1999. 'Why Should I Do Vaginal Examinations?' *The Practising Midwife* 2(6): 12–13.

WHO. 1994. 'Preventing Prolonged Labour: A Practical Guide. The Partograph 1: Principles and Strategy'. Geneva: World Health Organisation.

Winter, C. 2002. 'Orderly Chaos: How do Independent Midwives Assess the Progress of Labour', Masters dissertation. London: South Bank University.

Winter, C. and J. Cameron. 2006. 'The "Stages" Model of Labour: Deconstructing the Myth', *British Journal of Midwifery* 14(8): 454–56.

Zak P. J., R. Kurzban and W.T. Matzner. 2004. 'The Neurobiology of Trust', *Annals of the New York Academy of Sciences* 1032(1): 224–27.

# CHAPTER 5

# TIME AND MIDWIFERY PRACTICE

*Trudy Stevens*

To practice the science of medicine and analyse and treat the disease the physician distances himself or herself in time from the patient and treats the patient as allochronic, in another time.... To practise the art of healing the physician meets the sufferer in his or her own time, as a coeval.

—Frankenberg (1992: 10–11)

## Introduction

The idea that the practice of midwifery is both an art and a science has long been promoted, as demonstrated by the title of a textbook written for advanced midwifery practitioners (Silverton 1993). However, in the above quote, Frankenberg has suggested that the practice of science and the art of healing involve radically different approaches defined by different notions of time, approaches so different as to be distinct and separate.

In the arena of childbirth the distinction that Frankenberg drew between medicine and healing might readily reflect the ideological difference between obstetrics and midwifery; the one focuses on real or potential problems, whilst the other supports a physiological process. Therein lies one of the major tensions in current midwifery practice. The remit of midwifery is the promotion of normal childbirth, with practitioners being upheld as its 'experts' or 'guardians', and the concept of salutogenesis being adopted as a theoretical frame for the development of midwifery knowledge (Downe and McCourt 2008). Nevertheless, the push towards the hospitalization of birth for reasons of safety (DHSS 1970), and more recently the imperative to embrace the hegemony of evidence-based medicine (Enkin 2006; NMC 2008) has served to strengthen the 'science of midwifery', to the detriment of its art.[1]

Such issues emerged as a major theme in a doctoral ethnographic study of caseload midwifery, a radically different form of midwifery practice implemented within a National Health Service (NHS) maternity service in England. In this chapter I focus on how issues to do with time were seen to influence the nature of midwifery practice and how dramatically this impacted on care provision. Caseload midwifery effectively denied Frankenberg's thesis by embracing both the art and science of midwifery practice but the tensions that were generated, both within the organization and at a personal level, were seen to have their roots within conflicting notions of time.

## Background

Recognition of public dissatisfaction with NHS maternity services and the centralization of birth led to a major review concerning the maternity services in England and Wales (HoC 1992). Women had felt as if they were subjected to a conveyer-belt system, whilst midwives found that, despite their ethos of autonomy, the reality of their work was domination by the medical model of care. The recommendations of The Expert Maternity Group (DoH 1993), accepted as government policy in 1994, promoted the normality of birth and sought mechanisms for supporting this model, including the recognition of midwives as the appropriate main care providers.

Caseload midwifery practice was developed as a form of care delivery that implemented these recommendations. Although similar to the way independent midwives worked, this was a radically new organization of midwifery practice within the NHS, and the ramifications of implementing it within an existing maternity service were unknown.[2] In this model, each midwife, instead of being allocated to an area of the service, such as a hospital ward, cared for an annual caseload of forty women; this included care and education throughout the antenatal period, assistance with the birth, and subsequent care of mother and baby up until twenty-eight days postnatally (Stevens 2003). Rather than being based within a central hospital, the midwives worked with the women, where and whenever required. This necessitated them being available twenty-four hours a day and working in a range of different environments, including women's homes, general practitioners' surgeries and the maternity hospital. The use of mobile phones, and the avoidance of being tied to work at specific times, such as is the case with a clinic, facilitated such flexibility and enabled the midwives to 'make the job work for them'. Being organized into partnerships within groups of six to eight full-time equivalent midwives enabled the midwives to support each other and provide cover for 'timeout', holidays and sickness.

In this model, midwives were being given greater personal responsibility than they had previously experienced within the NHS, although no more than they were trained for and which the ethos of midwifery supported. Concerns were raised that midwives would be both unable and unwilling to work this way. The partnership and group structure were mechanisms by which support, cover and peer review of practice were facilitated but would this be enough to enable the midwives to function effectively, to work safely and efficiently and not become overtired, stressed or burnt out? A robust evaluation was built into this service development, one arm of which sought to explore and identify the implications this held for the professionals (Stevens 2003).

## Ethnography as Evaluation

Ethnography has not been an approach commonly adopted for evaluative studies: it has been seen to take too long and the findings are frequently framed as thick descriptions that enrich an understanding of a situation but preclude any quick answers. However, whilst other parts of the evaluation sought answers through more traditional quantitative and qualitative methods, it was recognized that exploring the implications for professionals was unknown territory. Any form of positivist enquiry would invariably frame the responses and limit the opportunity to understand the situation from the perspective of those being studied. An ethnographic approach was thus adopted as being most suited to the needs of this situation. Good ethnography, in terms of its validity, can only be achieved when the participants feel confident enough to behave in their usual manner and respond openly and honestly to the probing queries of the researcher. Thus it is advantageous if the researcher can minimize their impact on both the study setting and the participants, and is perceived as presenting no threat to the participants; such criteria are met when the researcher has experience of working within the field but is unknown to the study site, as in practitioner-researcher situations such as this study.

However, practitioner–research has been viewed with scepticism, thought of as entailing an inherent subjectivity because the researcher is unable to theoretically disentangle themselves from their work (Field 1991). Also, maintaining research awareness within a familiar setting and not inadvertently imposing their own 'world-view' are inherent difficulties which demand constant reflexivity from an ethnographer. This is comparable to the debate within anthropology about 'research at home' (Lipson 1991). Hammersley (1992) pointed out that the self-knowledge demanded of all ethnographers is not immediately given, and that people can deceive themselves and may even have an interest in self-deception. In being appointed to undertake this ethnography, my experience as a practising midwife meant that, to the maternity services, I was an 'insider', familiar with the setting, jargon and expected behaviour. However, long-term overseas experience, working with and for people who held very different views to myself, had forced me to confront my own views, assumptions and training. These experiences proved central in achieving the 'anthropologically strange' stance advised by Hammersley and Atkinson (1995: 9). Serendipitous circumstances allowed fieldwork to continue over a period of almost four years, which facilitated an understanding of both the initial implementation phase and its subsequent development into a more settled service. During this time, throughout the week I lived on-site in the 'nurses' home', so was able to participate in the life of the hospital twenty-four hours a day. I also worked for two days per week as a clinical midwife on the Delivery Unit, which sensitized me to the culture of the maternity service and many of the issues experienced by hospital midwives.

Data collection and analysis were undertaken in an iterative process: in response to the issues which emerged a range of data collection methods were used, including individual and focus group interviews with each of the key groups involved, where I sought to understand their unique perspectives, observation of practice, survey questionnaires and some documentary analysis. The findings from each group were considered in relation to each other and to the wider service delivery. It was at this stage of the analysis that the issue of time emerged as a dominant theme and the ways in which it controlled both the quality of care and midwifery practice became apparent.

Frankenberg's (1992) quote aptly highlights a fundamental difference between institutional birth and that facilitated by caseload midwifery practice. This chapter explores the different approaches to time that were observed within the hospital and caseload practice, and develops an understanding of how issues concerning 'time' were used as mechanisms for controlling childbirth. Such perspectives, although found to be fundamental to the nature of midwifery practice, are deeply embedded in the social life of the service and are unlikely to be tapped by positivist inquiry. This highlights the value of the ethnographic approach, enabling identification and exploration of this theme.

## Concepts and Uses of Time

Time is often thought to be a universal concept, one of the few immutable truths that help provide stability in an increasingly complex world. Nevertheless, many writers have shown this assumption to be fundamentally incorrect (Thompson 1967; Whitrow 1989; Priestley 1964; Hall 1959). Diverse notions about time have been identified, and the ways it is constructed, used and interpreted may hold widely differing connotations, both between and within societies (Griffiths 1999). The ways in which time is conceptualized and used can communicate powerful messages. In English, time has been externalized, made tangible, a commodity that can be 'bought' and 'sold', 'saved', 'measured', 'wasted', or 'lost'. It is compartmentalized, allocated for work, leisure and sleep, and it is used sequentially; it is valued objectively and personally, carefully guarded, and individuals becoming angry if 'their' time is unnecessarily wasted (Hall 1959, 1976), ideas that, it will be seen, are interwoven within hospital work. An understanding of how time was conceived within the hospital and within caseload practice reveals underlying notions that influence the nature of the services provided. However, as both were situated within the *durée* (Giddens 1987) of daily life, this must first be addressed.

The way time is conceived of and used in modern society has been strongly shaped by the influences of religion and technology. Judaeo-Christian beliefs stress the notion of irreversible time; 'switched on' at creation and to be 'turned off' in the future. Meanwhile, the Protestant work ethic (Weber 1976), placed a high value on the industrious use of time for spiritual rather than material rewards. Such notions, reinforced by puritan preachers and social reformers, were subsequently internalized during the Victorian era (Thompson 1967), promoted with the 'professionalization' of midwifery (Heagerty 1997), and remain in the ideas of some that nursing and midwifery are vocational work. As discussed in Chapter 1 (this volume), the industrial revolution had a profound effect, with time's 'inexorable passage' being stressed by mechanization that altered the rhythm of people's lives, negating seasonal or cosmological distinctions of time and reducing the element of personal control over work. The need for the synchronization of labour meant increasing attention was given to time, with people being paid by the hour not the task. Work itself became a distinct period of time, and time became a currency not to be 'passed' but 'spent' (Thompson 1967). Today, universal education inculcates a time discipline on all. 'Economic' time tends to dominate life, patterning its stages through infancy, learning, earning, retirement, each year (work and holidays) and each day, clearly dividing it into work and personal time – mentally if not physically. Diaries are no longer used to record events but to remind and structure them. The upsurge in the use of the Filofax and personal organizers, and development of various training courses, suggests that 'time management' has become an economy in itself. However, such concepts and their consequences are not universal and less industrial societies have been shown to hold very different notions of time. For all practical purposes 'task-orientated' time is the major framework (Giddens 1987; Priestley 1964); work is adjusted to the task not the time spent, and there is minimal demarcation between labour and social activities.

Although occurring in societies dominated by culturally specific notions about time, childbirth carries its own (universal) time – a physiological time. The mother commonly 'slows up' towards the end of pregnancy and may experience changes in sleep patterns. To a greater or lesser extent the expectant mother is being eased into having to use her time in a different way to meet the demands of a newborn that has yet to be socialized into a 'daily routine'. Labour commences with no reference to what may be socially convenient, and the woman is delivered into motherhood at a pace over which she has minimal conscious control. For millennia, 'traditional' birth attendants have supported and accompanied women during this transition, rarely attempting to control or subvert the timing of events that were physiologically inherent. This situation has changed radically in many societies (Davis-Floyd and Sargent 1997). In an age where time has become inherently scheduled and commodified, it is not surprising to find such control being extended to the arena in which childbirth is now placed.

Ideas about time are not homogenous to a society as individuals may favour particular notions. Also, in complex post-industrial society, people move between models during their daily life, being forced to acknowledge different attitudes and concepts relating to time simultaneously. For example: the demands for strict time control placed on factory workers contrasts with the generally more relaxed demands of family life; a similar difference was noted in my study within the hospital, between delivery unit and maternity ward. However, the dominant ideas become embedded within the culture of each society, both reflecting and influencing the ways in which people think and behave. This may have serious ramifications as concepts about time are relative to societies, dictating how individuals conceive their world and relate to each other. Problems occur when different sets of ideas about time clash – as when individuals move between countries or, as it is argued here, models of midwifery practice – forming the basis for 'cross-cultural' misunderstandings.

The ways in which ideas about time and its use can be internalized and affect behaviour have been most clearly developed by Hall (1959, 1969, 1976) and are helpful in understanding the different nature of caseload and hospital midwifery practice. Drawing from a number of disciplines, theoretical stances and empirical studies, Hall considered the notion of time and the ways this may influence a society. Using a comparative framework, he developed a thesis suggesting that time is not only a 'silent language ... speaking more plainly than words' (1959: 23), as well as something which structures behaviour and judgements made about that behaviour (1969), but it also influences cognition and the manner in which societies relate to their physical world (1976). His ideas offer invaluable insights into ways of considering social situations. For example, the 'task-orientated time' of pre-industrial societies is closely related to Hall's notion of polychronic time. This is characterized by several things happening at once and Hall stresses the involvement of people rather than adherence to pre-set schedules (1976). These characteristics may be seen to apply to caseload midwifery.

Modern post-industrial ideas of time are summed up in his notion of monochronic time, and Hall (1976) stressed how use of this directly affects attitudes and behaviour. Undertaking activities separately and sequentially implies implicit and explicit scheduling. This involves according priority to people and functions, and so forms a classificatory system ordering life which is so integrated that it appears logical and natural, although it is not inherent in natural rhythms. Prioritization implies a valuation, and thus the use of time acquires an implicitly recognized code: for example, a call at 2 A.M. has more serious connotations than one at 2 P.M. The segregation of activities enables total concentration but 'decontextualizes' them and people may become disorientated if they undertake several activities at once. Relationships are intensified but then temporally limited, as in business meetings or hospital appointments, which are private but of fixed duration. Failure to observe time limits implies an intrusion on another's schedule, and may be considered ill mannered or egocentric. Such ideas resonate strongly with the hospital maternity service and help explain negative reactions observed in my study towards caseload practitioners who worked within a polychronic timeframe.

In appreciating the changes faced by the caseload practitioners, an understanding of the way time was conceived and used within the hospital is important. Having come from this system the midwives would have internalized it to some extent. However, they were forced to rethink and develop different ways of using time in caseload practice.

## Hospital Time: An Uneasy Alliance

Implications concerning the way time and space are used and controlled within institutions like hospitals have been highlighted by studies such as Frankenberg (1992), Foucault (1973), Goffman (1968) and, in particular, Zerubavel (1979). A predominant feature of such work is an appreciation of the relationship between the control of time and status and power within the institution. For Frankenberg (1992), time itself and the way it was used and controlled formed a definitive element in the practice of healthcare and healing. Even though the majority of clients in maternity care are healthy women who could give birth successfully without medical intervention, it is managed institutionally within such a system. This relationship may hold particular implications for a maternity service that has been directed to provide mothers with increased choice and control in the care they receive (DOH 1994). How then was time used by the maternity service in this study and in what ways did the new model of care influence the caseload practitioners' ability to practise the art and science of midwifery?

The hospital maternity service necessitates the merger of three, potentially competing, time frames: physiological time, institutional time and the personal time of 'normal' daily life:

- Serving the needs of childbearing women, the *raison d'être* of the service is guided by the physiological time of gestation, labour and the demands of the neonate. The service has to be constantly available, twenty-four hours a day, 365 days a year.
- Serving the needs of many rather than the individual forces a rationalization and consequently the development of 'institutional time', which is described in this chapter.

- The service is provided by, and for, individuals who live in a world external to the hospital, and whose personal time is governed by the complexities of 'normal daily life' and the notions of time described previously. Work in, or visiting, the hospital is but one component in their lives.

In this study it became apparent that within the hospital these time frames formed an uneasy alliance, resulting in a particular patterning to the day and to the organization of work within it. Whilst tensions between physiological and institutional time were most apparent on the delivery unit, the potential for conflict between institutional and personal time occurred throughout the hospital, in all departments and wards. Although core staff working rotational duties or 'shift work' provided the twenty-four hour baseline service, institutional time gave the appearance of the patterning of activities of 'normal daily life'. Most categories of staff worked a modified 'office hours' regime; afternoon and evening visiting gave a social element to the day; and night time was a period of quiet, reduction in noise and lighting being used to encourage 'patients' to rest. Nevertheless, it could be extremely busy at night, and a reversal of the natural day/night, work/sleep dichotomy was imposed by bright lights being kept on. This subversion of 'normal-daily-life' time by institutional time was not remarked on by staff and generally accepted by 'patients'. Time was less tightly controlled over weekends and bank holidays when routine work was avoided and a more relaxed atmosphere prevailed.

The division of time and labour in the shift pattern of work was aimed at ensuring an appropriate number of appropriately skilled staff was available when most required, although this does not succeed in practice since labour and birth cannot be scheduled in the manner of work shifts (Audit Commission 1997). Additionally, a clearly hierarchical pattern was discernable. The association of flexibility and control over one's time being inversely related to status and power within a hospital has been highlighted by Zerubavel (1979), and was similarly noticeable here. Night periods were covered by more junior staff, supported by senior or specialist staff working an on-call system; the most senior staff, consultants and managers, were rarely seen at night unless called specifically for an emergency situation – their presence indicated that something was seriously amiss.

Although notionally serving the needs of twenty-four hour physiological time, hospital time imposed a strict schedule. The day was divided and defined by the clock in the organization of duty rotas, clinic schedules and appointments, ward rounds, operation lists and in-patient meal times. These determined where people should be at specific times of each day and helped ensure all necessary tasks were undertaken. In this way, time served to regulate and create order out of complex and, given the numbers of people involved, potentially chaotic situations. Adherence to these 'demands' generated the impression of efficiency and organization, even though it was not possible in practice to match staff levels in a shift system with the less predictable patterns of labour and birth.

The midwives themselves noted how different perceptions of time dominated different departments within the hospital. The Outpatient Clinic comprised two three-hour, sharp bursts of intense activity each day. These fitted relatively easily into the 'normal-daily-life' time of staff and attendees, acknowledgement of which was emphasized by the importance placed on punctuality, highlighted by waiting-time audits, even though in practice women waited long periods for very time-limited visits. The in-patient wards attempted to establish a 'normal-daily-life', 'physiological time' twenty-four hour rhythm to the day, although this was moderated by ward routines, set meal times, rest times and the regulated social contact of restricted visiting times. It was also sharply divided by the fast turnover of admissions and discharges, and the accompanying administration created intense work pressure for staff even though this was of a relatively non-urgent nature.

Perceptions of time, and the way it was used, proved very different on the Delivery Unit, where all births took place; it was here that the potential for conflict was most apparent. Providing a constant level of cover, the difference between night and day was appreciable only by a reduction in the number of staff seen on the unit. The use of bright lighting, particularly when busy, defied diurnal variations. However, physiological time cannot be overruled with the same ease, and inter-professional conflicts of understanding and approach around this emerged as the 'active management' of obstetrics versus the 'waiting' of midwifery (see also chapters 3 and 4, this volume).

To some extent the timing of work was initiated and ordered by physiological time – such as the spontaneous onset of labour – although institutional time was superimposed with work created by elective caesarean sections and inductions of labour. However, it was rare for physiological time to be allowed to proceed without some element of control. Even physiological labours progressing 'efficiently' and 'normally' were monitored by the clock; constant assessment of contractions in terms of frequency and duration, routine monitoring of the fetal heart, and regular assessments of progress helped tie the labour to chronological time. This was reinforced by a formal, supposedly research-based time frame imposed on the process of labour through the use of the partogram (Rosser 1994; see also Chapter 3, this volume). Further control of physiological time was both symbolized and actuated by 'the board' in the Delivery Unit office

In common with other maternity hospitals in England (Hunt and Symonds 1995), 'the board' contained the basic information relating to each woman admitted to the Delivery Unit; as such it provided a visual representation of the current clinical workload of the unit. It was the responsibility of the midwife caring for a woman to update the board as appropriate. This enabled the obstetricians and the sister in charge of the unit to be kept fully informed of an individual's condition, particularly the progress of her labour, without disturbing the mother or midwife caring for her. A report of a 'delay in progress' on the board would be watched carefully by the obstetricians who then proactively involved themselves in care management, before the midwife called for assistance.

Thus, the board, or more specifically the interpretation of the information presented on it, was seen to have a direct impact on behaviour and the subsequent workload of the unit. In many ways it provided a lynchpin for the working of the unit and a medicalized, 'management' approach to labour. This was symbolized in the information that was considered to be relevant to the board, and actuated through the 'progress reminders' it constantly presented. As such, the board became the focus for some tension between caseload midwives and the obstetricians and the sister, particularly when the midwives failed to maintain the information on the board, or to behave, as expected.

> The doctor came in and was looking down to see who was fully[3], who was pushing and who wasn't pushing and why not – and noticed that someone had been fully for a good length of time and why hadn't they delivered? (hospital midwife interview)

The controlling influence of the board was clearly recognized and frequently subverted by some of the hospital midwives, although the subversion tended to remain hidden and so did not challenge the established ordering of events. The unique physiological timing of a woman's labour may differ from the guidelines established by the authoritative knowledge (Jordan 1993) defining 'safe' limitations to the stages of labour. Noting events, such as the start of the second stage, on the board, 'sets the clock ticking' (a term used by midwives and mothers alike) and a mother not delivered within the allocated time would soon receive medical assessment. However, some midwives prevented such interventions by delaying tactics that avoided 'starting the clock' by, for example, not confirming the start of second stage when suspected and if maternal and fetal well-being were assured.

Many of the midwives complained about obstetricians watching the progress of labour too assiduously, being too interventionist and expecting women to be examined vaginally at two-hour intervals so as to monitor progress. Such regularity was not indicated in the procedures manual nor, in personal experience, imposed by the obstetricians. However, during personal clinical practice, an experienced sister advised me to undertake such regular examination 'as the doctors expected it', a situation also experienced by others:

> Here, if the doctor doesn't come and knock, in two hours the sisters will – they are pushing the doctor to ask how things are progressing. To get a breather I give in. OK, come and knock. (focus group, hospital midwives)

In the hegemonic medical model, labour is not a safe time for mother or baby, and judicious intervention is indicated when there is a delay in the process. Although disputes over what constituted 'delay' were recognized, medical guidelines concerning appropriate time frames were followed. Perceived delays in progress were quickly noted and intervention was recommended, a system not just dependent on obstetricians' actions but internalized and practised by senior midwives.

The possibility of complications encouraged an immediate time orientation and it was recognized that the pace of work on the unit would, at times, quickly change. As one midwife commented 'they work in hours down there' referring to the wards 'whilst we work in minutes up here!' The peaks and troughs of work that are inherent in childbirth and the maternity service generate a clash between the rhythms of nature and those of the institution. At times staff had to remain on duty when there was little work to do; at other times the pressure of work was so relentless and staff so limited they quickly became exhausted and worried about safety levels being compromised. A seemingly constant fear of litigation served to increase the stress of these periods.

### Implications for Midwives and Midwifery

In providing a twenty-four hour service to a large number of women, the institution developed a momentum of its own. This seemed to have an inherent logic to it, which was then internalized and reinforced by the staff. In accepting employment, hospital midwives gave complete control over the timing of their work to their employers, who set the shifts and rotas on a three-week cycle and thus exercised a high degree of control over their personal lives. Midwives could submit requests for particular duties to their employees but these were not necessarily granted; a few subverted this control by occasionally reporting sick when a requested day off had been refused. Acknowledging the Sapir-Whorf hypothesis (Sapir 1985; Whorf 1971), the accepted use of the term 'days off', rather than 'days on', linguistically reflected the domination institutional time had over the midwives' personal time. Personal life was arranged around the needs of the hospital, often to the detriment of the individual – particularly those with young children – as witnessed in tensions generated over cover scheduled for school holidays, Christmas and New Year. The majority of midwives grumbled about personal difficulties incurred but appeared to accept this as 'part of the job'. Institutional time was accepted as the 'norm' for midwifery work.

Not only did the hospital midwives have very little influence over when they actually worked, whilst at work they had minimal control over the place and content of their working time. Meal breaks were taken when allocated rather than chosen to suit the workload situation; not infrequently in the Delivery Unit, the relentless demands of crisis situations precluded meal, coffee and even toilet breaks. Although Hall (1959, 1976) described notions of 'modern' time as being scheduled and prioritized, within the hospital the midwives were frequently required to undertake many tasks at once, juggling the competing demands of a busy unit, incessant telephone calls, crying babies, concerned relatives and clinical emergencies. Not in ultimate control of such situations, the midwives were forced to be reactive rather than proactive and exhibited the disorientation identified by Hall (1969).

The tightly defined boundaries of the midwives' time generated a short-term focus that forced them into an immediate-task orientation, akin to a Fordist division of labour (Godelier 1988) where activities are broken down to their component parts and undertaken separately. The rotational nature of midwives' duties limited the possibility of them caring for the same woman on their next shift; thus, it became almost irrelevant for the midwives to develop an understanding of the mother's situation – the wider context of the care they provided. The philosophy of continuity of care (similar care provided by all staff) was acknowledged as being ideal, but so was the reality of conflicting advice given by different professionals.

Given the relatively short duty span in the context of longer care requirements, midwives were unlikely to complete care provision; they had to leave when it was time to go off duty rather than stay and complete the activity, such as assisting with a birth. Thus time divisions, rather than completion of a task, becomes the guiding focus of work. This did not sit comfortably with the midwives and many would 'stay behind', or miss meal breaks, even when a relief was available, if this was at an inappropriate time for the mother. However, such practices were not encouraged. For example, one midwife reported how a sister 'refused to allow' her to stay on duty for the delivery of a mother she had been looking after all evening because she was expected back on duty the next morning.

Hospital midwives were contracted to work 37–and-a-half hours per week with a specific holiday entitlement. Payment for extra hours worked was not available, except in exceptional circumstances, and midwives were expected to 'take back' time when the unit was quiet by going off duty early. However, the reality of understaffing and increased workloads meant that they were rarely able to do this. Several senior midwives were 'owed' many hours, which they recognized they would never be compensated for. True commoditization of their time had failed, ironically resulting in the institution 'stealing' an employee's time because they had focused on completing the activity for which they were employed rather than the time 'allowed'. The use of time within the maternity hospital took on symbolic valuation and, most importantly, developed a momentum that appeared unalterable. Scheduled time became predominant, internalized and accepted as the normal, sensible way of 'doing things'. This held important implications for the way midwifery care was delivered and for the midwives as individuals. Such notions were challenged by caseload midwifery practice.

## Time and Caseload Midwifery

Caseload midwifery practice (CMP) required a radically different orientation towards time. The new style of practice challenged the notions previously developed within the hospital service, forcing midwives to redefine their concepts about time and its use. In 'giving back' control over time to midwives, the maternity service implicitly acknowledged the control it exercised over those remaining in the conventional service, a feature that was apparently not overtly recognized. The different orientation towards the use of the caseload midwives' time was structurally defined within their contract. They were employed to undertake broadly specified activities (or responsibilities) rather than provide a set number of midwifery-care hours. Operationalization of this requirement was at the discretion of the individual midwife, and fixed additional payment, irrespective of actual 'unsocial' hours worked, facilitated their flexibility. This strategy effectively de-commodified the midwives' time. It also removed the pressure to complete an activity within a specific time, such as before going off duty.

| Hospital Midwives | Caseload Midwives |
|---|---|
| Contracted for 37½ hours work per week | Contracted for care of 40 women per year |
| Commodified time – extra payment for 'unsociable hours' | Set extra allowance irrespective of time of day worked |
| Extra hours worked not paid | Not applicable |
| Clear divide between work and personal life | Work 'embedded' in personal life |
| Request particular days off | Negotiate free time with partner and group, or by managing own workload |
| Minimal flexibility to change duty | High level of flexibility |
| Work according to fixed duty rota | Work when needed by women |
| Work period intensely busy or quiet. Unable to take advantage of quiet periods. No balance reported. | 'Long hauls' and quiet periods when minimal work. Can use to personal advantage. Reported to balance over time. |
| Work 'time' directed and controlled by hierarchy | Self-directed except where 'controlled' by labour and emergencies |
| Rota orientation – leave work when 'due off' – obstacles to staying | Activity orientation – finish work when activity completed |
| Current work has present orientation (task in hand) | Current work has future orientation (investment in future care provision) |
| Midwives' 'time' has a future orientation – immediate future work time known | Midwives' 'time' has present orientation – immediate future work time uncertain |
| **Time is routinized, controlled, scheduled, de-personalized** | **Time is purposeful, flexible, uncertain, personalized** |

Figure 5.1: A comparison of orientations towards, and use of, time for midwives.

By altering the focus of work from time to activity, midwives worked when and as they determined or were required. Thus, although strict time scheduling of work is often associated with 'efficiency', they were able to use their time more effectively, no longer having to 'waste' it by going 'on duty' when it was quiet and no work was actually required, by having to duplicate work or to handover in mid activity owing to shift changes. Without close managerial direction, the midwives now 'owned' their time and were able to deploy it as they considered appropriate, spending as long or as short a time as they considered appropriate to achieve the activity in hand. One midwife, describing how she managed this situation, noted: 'I tend to do less visits over a longer time', that is, visits were of a longer duration. This presented them with enormous flexibility. Inevitably some variation in the way they structured their time developed. Some chose to start work early, others later in the day; some scheduled their routine work into a few long days whilst others planned for a more even spread.

Arranging cover at night and weekends was equally flexible. Some midwives preferred to remain available for their women, recognizing the limited chance of being called, whilst others opted for alternating night cover with their midwifery partner, preferring the higher chance of being called one night with the certainty of not being disturbed the other. Such flexibility enabled each midwife to negotiate with their partner a pattern of working that best suited their lifestyle. Moreover, as their lives and commitments changed, such patterns were relatively easy to alter and adapt.

> You actually have to plan better when you are working shifts. I find I plan on a weekly basis. Whereas before, when I was on the wards, you have to plan three weeks in advance because that's the way the rotas are done. (interview caseload midwife)

The midwives did not have total control over their time as they had to be available to respond to the needs of their women. Nevertheless, once they had developed their personal time-management skills and learnt to advise, or 'educate', their women appropriately they reported that interruptions at night were usually confined to labour and emergencies and proved to be minimal.

> At night? It's not very often. I would say on average a month I would get three. You can't put (a number on it). Or you may be contacted three times in one night! (interview caseload midwife)

Such reporting was verified in a study of the midwives' work diaries (McCourt 1998). Knowing the women who contacted them enabled the midwives to respond appropriately, not necessarily having to make visits but instead give advice or make an appointment. This contrasted with their colleagues in the conventional services where calls from 'unknown' women had to be treated with care; with no prior knowledge of particular situations, most calls necessitated the woman being asked to come into hospital or being visited at home by the community midwives.

These two features, knowing the women and infrequent night calls, were symbiotic. Relating to their caseload midwife as a person rather than as a role, women were reported as saying they did not want to disturb them unless it was urgent. This appeared to be one of the most misunderstood features of caseload practice. In considering this model of care, both midwives and doctors understood the term 'on call' in terms of the hospital system in which, in their own experience, they were invariably disturbed. Alternative models, where they were 'available' yet rarely called, appeared incomprehensible.

As their time was not tightly defined or structured, and largely under their control, the caseload midwives were able to work within women's individual time constraints and their physiological time frame. With minimal previous experience of home births, the midwives reported finding that deliveries at home had a very different quality. They became more aware of the physiological rhythms of labour which, away from the constraints of hospital dominated time, were found to be very different from what they had previously considered as 'normal'. The midwives considered they learnt this by having to advise women during the early stages of labour and then caring for them through the active phase, rather than providing an eight-hour period of care isolated from the wider context of labour.

The caseload midwives tried to subvert the hospital time imposed on labour by their own strategic use of 'the board' in the Delivery Unit; as previously noted, this (open) refusal to comply with accepted procedure generated tension on the unit. Also, with a greater understanding of individual situations, they became more flexible in applying the unit's guidelines and protocols concerning labour. In describing a difficult delivery involving a long second stage, one caseload midwife explained that, because she was aware that the mother was unsure of the parentage of her child and was fearful of her baby's colour at delivery, she considered the delay was due to the mother psychologically holding back. In this situation the midwife considered that, while indications of the baby's well-being were satisfactory, support and understanding were more appropriate than speeding up the labour with hormonal stimulation.

### Implications for Caseload Midwives

Such flexibility held distinct advantages for midwifery practice and mothers. Nevertheless, personal adaptation by the midwife was not necessarily easy or successful. This study indicated that it took between six to twelve months for midwives moving from the hospital service to settle into working this way, and that the most fundamental adaptation, although not overtly recognized, was likely to be to different notions and uses of time. Their lives were no longer clearly compartmentalized into the scheduled, tripartite divisions of Hall's (1969) monochronic time – work, social and domestic time, and sleep – as work became instead embedded in the general passage of their lives in much the way Bourdieu (1963) described for Algerian peasants and Bohannan (1967) argued for the Nigerian Tiv. This lack of the compartmentalization of time may also be considered a feature of postmodernity, with the movement to more flexible patterns of working, in both time and space, indicated by the development of 'flexi-time' and home-offices. It is certainly a feature of the lives of some of those in more autonomous positions, such as senior corporate managers and senior professionals (Giddens 1987), and this was perhaps linked to the greater professional confidence and respect from obstetricians that the midwives began to experience (Frankenberg 1992).

This way of using time had a direct impact on the way the midwives viewed their lives, but it also held a certain ambiguity. Long-term planning was important for negotiating holiday time, and a balance to the caseload; it also incorporated the essence of 'investment' in their work discussed previously. However, short-term planning was less assured, forcing a more 'present' orientation and a need to be able to live with uncertainty. Although they would know 'due dates' for delivery and might have a sense of impending labours, they never knew when they would be called. Even when quiet, their busier colleagues might require support. The midwives recognized these patterns balanced out, that periods of intense activity would be followed by quiet spells. However, their appreciation of the quiet times was probably more retrospective than immediate, the exact duration of the quiet period only being defined once it had passed.

On a day-to-day basis the development of a forward orientation was limited as anything planned during 'available' periods could be disrupted by unexpected labours or emergencies. The ability to plan in certainty and enjoy the anticipation of particular social activities was determined by the support provided by their partners or group, and defined by whatever strategies for cover they had negotiated. The midwives' mobile phones became both the symbol and reality of this embedded work, freeing them to go wherever they wanted, as far as was reasonable and socially acceptable for the use of mobile phones, when they were officially 'available' but also interrupting such activities with the demands of their caseload. This extended into all aspects of their lives, with *coitus interruptus* being described laughingly by some as a new form of contraception. Adaptation to this 'embedded' more 'traditional' use of time was dependent on both personal characteristics and personal situation. It clearly suited those with a flexible and relaxed attitude towards work and life in general, proving more problematic for those who enjoyed living very structured lives. This different approach to 'work time' also made different physical demands on the midwives.

### Time Clashes

Many of the difficulties the midwives experienced as caseload practitioners related to clashes experienced at the interface between their 'traditional', 'postmodern' concepts and uses of time and others' 'institutional' or 'modern' notions. These occurred in their domestic lives, with some of their clients, and when working in the hospital. Clashes that developed in the domestic domain were highly individual, and depended on particular circumstances. Being called when socializing with friends was difficult for some, whilst others said they experienced minimal problems in negotiating such situations. Most midwives commented on not being able to drink alcohol when they were 'available' to be called by women, but reported adapting to this. Midwives with stable and established live-in relationships appeared to experience less domestic tension than those with new or changing relationships. The greatest problems occurred when couples lived apart, particularly if separated by any distance. Tensions arose when visits together were interrupted by calls to work.

Two midwives reported finding childcare when working with a caseload considerably easier than with the shift pattern of work, but they acknowledged they benefited from flexible and supportive domestic arrangements such as the close proximity of supportive 'grandparents'. Others experienced greater difficulty, and reported feeling guilty when relying on friends to assist. This situation exemplifies one of the difficulties of using time in a more traditional way within a society that is structured and dominated by scheduled, industrial time. In traditional societies, childcare is commonly conceived of as the responsibility of the wider family, not just the mother. Where specialized childcare arrangements have to be adopted the uncertain nature of caseload practice can result in high fees or high levels of stress.

Although the reports were few, it became apparent that some clients experienced difficulty with the flexibility that was an integral part of the midwives' use of time. Living within a structured, scheduled time frame, their highly organized lives were disrupted when planned visits had to be cancelled at short notice (for example, when a midwife had to go and attend another mother's labour). One husband wished to lodge a formal complaint to the Health Trust, explaining how angry he had become when, having cleared time from his city occupation in order to meet the midwife, this visit was postponed at the last minute. He clearly considered his time had been 'stolen' by the midwife's inefficiency. In industrialized countries, punctuality is indicative of efficiency, although elsewhere aspects relating to respect, status or power are more heavily stressed (Hall 1959, 1976). Such clashes, unless recognized and tactfully handled, irritated clients who then interpreted the midwife's behaviour as disorganized or unreliable.

More serious difficulties developed when the midwives interfaced with the hospital service, where institutional time predominated. Problems were generated both in the way activities were undertaken and the negative stereotyping which developed from misunderstandings, a situation well recognized in cross-cultural misunderstanding relating to time (Hall 1959, 1969, 1976; Carroll 1990; Griffiths 1999). The interface in the Outpatient Clinic was reported as a constant problem by both groups of staff. The clinic was managed on a tight schedule and the hospital midwives reacted sharply when caseload midwives did not appear as arranged or spent 'too long' with women, 'blocking' rooms and disrupting the 'smooth running' of the clinic. In the more relaxed atmosphere of the inpatient wards, the hospital midwives still complained that the caseload midwives were inefficient and disorganized, lazy and poor time keepers; they appeared at irregular times of the day and could not be relied upon to attend when planned, descriptions not infrequently applied to the same hospital midwives by the caseload practitioners. Both students and junior midwives noted how some hospital midwives phoned the caseload midwife for non-emergency queries at any time of day or even night. The perception was that as hospital midwives' shifts covered the hospital twenty-four hours a day so did the caseload midwives, and so it was felt appropriate to contact them at 3 A.M. for a minor query.

In the Delivery Unit, where time took on a shorter, more concentrated dimension, the relaxed attitude and flexibility of the caseload midwives proved particularly irritating to hospital shift-based staff if the unit was busy as described in Box 1:

---

**8.30 A.M.**
The unit is frenetically busy, staffing is difficult and there are a number of emergencies. Access to the telephone is constantly required.

One of the two phones is being used by a caseload practitioner to reschedule her day's work, having brought in a lady in labour. She is unaware of the intense irritation she is generating by her relaxed and humorous, although totally work orientated, conversation. Her use of the phone lasted about ten minutes.

Nothing is said but strong 'looks' are exchanged between medical and midwifery staff.

**Note**: The caseload midwife's character was visually assassinated! A clear example of a 'time-clash'.

---

*Source* DU.observation study no.10 1997.

A second area of tension arose between the shorter periods of duty and longer duration of caring for a woman throughout labour, where caseload midwives received little help from hospital staff. Particularly in the early days, the hospital midwives considered it inappropriate to offer help, as the caseload midwives were responsible for their own caseload. However, they did not fully appreciate how long a particular caseload midwife had been on the unit, nor their previous workload prior to attending the labour. This withdrawal of support may have been fuelled by the caseload midwives' initial reluctance to 'update the board' in the unit (see above) because they did not wish to 'set the clock ticking' and end up being dominated by medical time and 'interference' unless they requested advice or support. As a result, obstetricians accused the sisters in charge of the unit of 'not knowing what was happening'. As a result, some of the sisters appeared to marginalize the caseload practitioners.

## Time and Radical Change

Frankenberg (1992:16) suggested that 'revolutionary changes in health services ... require that time itself is turned upside down', commenting how, in *Das Capital,* Marx exhorted workers to take charge of their own time. He also noted how a more egalitarian form of healthcare, defining carers and cared for as equal participants in the healing process, would neither need nor be able to treat the time of others as within its control. Practising with a caseload involved a radical change for midwives, not least in the way time was conceived and controlled, and this held fundamental implications for the midwives' work and lifestyle. The more reciprocal relationships established with mothers included mutual respect for each other's time and, with a less controlled patterning of their own time, midwives gained a greater appreciation of the physiological timing of labour.

Frankenberg (1992) remained pessimistic as to the viability of the change he had outlined, considering such relinquishment of power to be idealistic. Somewhat appositely he used the metaphor of childbirth when presenting this idea, suggesting that 'historical changes, like women in labour, still need midwives, even if for both they can most usefully be chosen from among their friends' (Frankenberg 1992:18). The nature of caseload midwifery practice appeared to support his views on revolution and egalitarian healthcare. As this study indicates, the fact that it has been successfully implemented, although only as a small scheme, and is subject to high levels of resistance and inter and intra-professional tension, undermines his pessimism but concurs with his valuation of 'friends', albeit 'professional friends'.


## Conclusion

In analysing the adaptations carrying a caseload demanded of the midwives, it was apparent that particular structures that had become separated in 'modern' society became fused again. The 'role' and 'person' of the midwife became one, and the professional/client dichotomy became a relationship of mutuality where the expertise of both midwife and mother were valued. Such fusion presented a radical alteration to the way caseload midwives worked. However, perhaps the most fundamental fusion they experienced related to their use of time. This necessitated a deconstruction of the 'modern' way of compartmentalizing time, returning to a more 'traditional' way of conceiving and using it (Thompson 1967). Frankenberg (1992) indicated that a different use of time was involved in the practice of the science or the art of 'curing'. So it was in caseload midwifery. The different way of using their time enabled midwives to meet mothers on a level that acknowledged and facilitated the physiological timing of childbirth. Nevertheless, this change conflicted with institutional concepts of time and the way time was used by others, generating tensions.

Ideas about time, and the expectations generated by these, influence the way people live and relate to others. This understanding of the way time was used, both within the hospital and when carrying a caseload, helps give an appreciation of the very radical differences between the two models of practice studied here. It may also help explain some of the problems experienced, by all groups of staff, in responding to this social change. Those that work in the maternity services are part of a wider as well as local social world, and the implications of such change were wider than the immediate work context. The changes in uses of time observed in this study also implicitly challenged the related issues of hierarchy and gender in caring work, and ways of managing labour and birth.

## Notes

1. There is a theoretical debate about the meaning of evidence, which is discussed in Downe, S. and C. McCourt, 2008. 'From Being to Becoming: Reconstructing Childbirth Knowledges', in S. Downe (ed.) *Normal Childbirth: Evidence and Debate*, 2$^{nd}$ edn. London: Churchill Livingstone, pp. 3–24. Here we refer to the relatively narrow definitions of 'science' and of 'evidence' that are used in practice in evidence-based medicine and regarded as authoritative knowledge. These hinge on viewing experimental research, particularly randomized controlled trials, and meta-analysis of trials, as the only really valid form of evidence. The concepts of evidence and science could be interpreted more widely, but this is not established in biomedicine.
2. Although described at the time as radically new, this was only within a short-term historical perspective. As discussed in Chapter 1 (this volume), this pattern of work was similar to that followed by most traditional midwives worldwide, and had been the norm for midwives in the U.K. until they became salaried employees with the advent of the NHS in 1948.
3. This is midwifery 'shorthand' for full dilatation of the cervix, which is considered to signify the beginning of the second stage of labour.

## References

Audit Commission. 1997. *First Class Delivery: Improving Maternity Services in England and Wales*. London: Audit Commission for Local Authorities and the NHS in England and Wales.

Bohannan, P. 1967. 'Concepts of Time Among the Tiv', in J. Middleton (ed.) *Myth and Cosmos*. New York: Natural History Press, pp. 315–29.

Bourdieu, P. 1963. 'The Attitude of the Algerian Peasant Towards Time', in J. Pitt-Rivers (ed.) *Mediterranean Countrymen*. Paris: Mouton, pp. 55–72.

Carroll, R. 1990. *Cultural Misunderstandings: The French–American Experience*. Chicago: University of Chicago Press.

Davis-Floyd, R.E. and C.F. Sargent (eds). 1997. *Childbirth and Authoritative Knowledge: Cross-Cultural Perspectives*. Berkeley: University of California Press.

DoH. 1993. 'Changing Childbirth', *Report of the Expert Maternity Group, Part 1 (Cumberlege Report)*. London: HMSO for the Department of Health.

DoH. 1994. 'Woman-Centred Maternity Services', *NHS Management Executive Letter EL(94)9*. London: Department of Health.

DHSS. 1970. *Domiciliary Midwifery and Maternity Bed Needs (Peel Report)*. London: HMSO for the Department of Health and Social Security.

Downe, S. and C. McCourt, 2008. 'From Being to Becoming: Reconstructing Childbirth Knowledges', in S. Downe (ed.) *Normal Childbirth: Evidence and Debate*, 2nd edn. London: Churchill Livingstone, pp. 3–24.

Enkin, M. 2006. 'Beyond Evidence: The Complexity of Maternity Care', *Birth* 33(4): 265–69.

Field, P. A. 1991. 'Doing Fieldwork in Your Own Culture', in J. Morse (ed.) *Qualitative Nursing Research: A Contemporary Dialogue*. London: Sage, pp. 91–104.

Foucault, M. 1973. *The Birth of the Clinic: An Archaeology of Medical Perception*. New York: Pantheon.

Frankenberg, R. 1992. 'Your Time or Mine: Temporal Contradictions of Biomedical Practice', in R. Frankenberg (ed.) *Time, Health and Medicine*. London: Sage, pp. 1–30.

Giddens, A. 1987. *Social Theory and Modern Sociology*. Cambridge: Polity.

Godelier, M. 1988. 'Foreword', in B. Doray, *From Taylorism to Fordism: A Rational Madness*. London: Free Association Press, pp. i–vi.

Goffman, E. 1968. *Asylums: Essays on the Social Situation of Mental Patients and Other Inmates*. Harmondsworth: Penguin.

Griffiths, J. 1999. *Pip Pip: A Sideways Look at Time*. London: Flamingo.

Hall, E.T. 1959. *The Silent Language*. New York: Doubleday.

——— 1969. *The Hidden Dimension*. New York: Anchor Books.

——— 1976. *Beyond Culture*. New York: Doubleday.

Hammersley, M. 1992. *What's Wrong with Ethnography*. London: Tavistock.

Hammersley, M. and P. Atkinson. 1995. *Ethnography Principles in Practice*, 2nd edn. London: Routledge.

Heagerty, B.V. 1997. 'Willing Handmaidens of Science? The Struggle Over the New Midwife in Early Twentieth-Century England', in M. Kirkham and E. Perkins (eds), *Reflections on Midwifery*. London: Bailliere, pp. 70–95.

HoC. 1992. *Maternity Services: Second Report to the Health Services Select Committee (Winterton Report)*. London: HMSO for the House of Commons.

Hunt, S. and S. Symonds. 1995. *The Social Meaning of Midwifery*. London: Macmillan.

Jordan, B. 1993. *Birth in Four Cultures: A Cross-cultural Investigation of Childbirth in Yucatan, Holland, Sweden and the United States*. Prospect Heights, Illinois: Waveland Press.

Lipson, J.G. 1991. 'The Use of Self in Ethnographic Research', in J. Morse (ed.) *Qualitative Nursing Research: A Contemporary Dialogue*. London: Sage, pp. 73–89.

McCourt, C. 1998. 'Working Patterns of Caseload Midwives: A Diary Analysis', *British Journal of Midwifery* 6(9): 580–85.

NMC. 2008. *Code of Professional Conduct*. London: Nursing and Midwifery Council.

Priestley, J.B. 1964. *Man and Time*. London: Aldus Books.

Rosser, J. 1994. 'World Health Organization Partograph in Management of Labour', *MIDIRS Midwifery Digest* 4(4): 436–37.

Sapir, E. 1985. *Selected Writings in Language, Culture and Personality*. Berkeley: University of California Press.

Stevens, T. 2003. Midwife to Mid Wíf: A Study of Caseload Midwifery. Ph.D. dissertaton. London: Thames Valley University.

Silverton, L. 1993. *The Art and Science of Midwifery*. London: Prentice Hall.

Thompson, E.P. 1967. 'Time, Work Discipline and Industrial Capitalism', *Past and Present* 38: 56–97.

Weber, M. 1976. *The Protestant Ethic and the Spirit of Capitalism*. London: Allen and Unwin.

Whitrow, G.J. 1989. *Time in History*. Oxford: Oxford University Press.

Whorf, B.J. 1971. *Language, Thought and Reality*. Cambridge, MA: MIT Press.

Zerubavel. 1979. *Patterns of Time in Hospital Life*. Chicago: University of Chicago Press.

## CHAPTER 6

# 'WAITING ON BIRTH': MANAGEMENT OF TIME AND PLACE IN A BIRTH CENTRE

*Denis Walsh*

### Introduction: Research Issues and Setting

This chapter discusses temporality in relation to birth-centre care and forms part of the findings of an ethnography of a free-standing birth centre (FSBC) in the United Kingdom. The background to the study was the increasing interest in birth centres as a way of reducing interventions in childbirth. Intervention rates are escalating across the Western world with some U.K. studies showing that only a small minority of women experiencing their first birth do so without recourse to common medical interventions like syntocinon to speed up labour or operative delivery to terminate it (Downe, McCormick and Beech 2001; Mead 2008). A number of authors have suggested that these spiralling rates of intervention are linked to the centralization of birth in larger and larger hospitals (Wagner 2001; Tew 1998) and have suggested that small-scale, midwifery-led facilities and increasing home birth provision may reduce the medicalization of labour.

In the debate around place of birth, there is another divide emerging about the status of knowledge associated with quantitative and qualitative research methods. Davis-Floyd and Davies (1997) have challenged the over reliance on quantitative methods which are seen to be authoritative and superior to evidence produced from other sources. They suggest that important knowledge flows from expertise, embodied birth experiences and qualitative research and that these need to be considered alongside quantitative outcomes. Ethnography, a research method from the qualitative paradigm, was used in the study of the birth centre discussed here. It was adopted because of its strengths in drawing out contextual features and nuances from the study environment. These are more helpful in exploring features of birth-centre care that impact on women's experience of care and rates of birth interventions.

A further critique of quantitative research challenges its claims to objective, 'factual' knowledge. Downe and McCourt (2008) explore the assumptions underpinning positivism, the theoretical position that quantitative enquiry is based on. They argue that notions of simplicity, linear cause and effect, and certainty are an inadequate foundation for knowledge claims about human experiences and phenomena in childbirth. Complexity theory may have more to offer in researching maternity care because it takes into account the connectivity and inter-relatedness of different aspects of the complex nature of childbirth. What would flow from this exploration are far more tentative knowledge claims and this resonates strongly with the postmodernist era that many writers believe Western societies are entering (Fox 1999; Holub 1992). The ethnographic study discussed here seeks to reflect this more relativistic reading of findings while arguing that the conclusions drawn offer a convincing interpretation of the data.

At first glance, the role of time in the three issues referred to here – the medicalization of childbirth, the high value placed on quantitative research methods, and the objectivity of knowledge claims – is straightforward. Increasing interventions in childbirth may be related to the policing of the time line of labour. This will be explored further below by critiquing assembly-line approaches imported to maternity care from industry. Time, it is claimed, has measurement as an intrinsic property and, therefore, it can be readily scrutinized by quantitative research methods. Finally, the objectivity of time seems axiomatic. It appears to exist outside culture and the social with both spheres being regulated by it. These ideas will be contested in the following sections.

From within the quantitative paradigm, research to date on FSBCs is widely acknowledged as very heterogeneous, making generalizations problematic, though all studies favour FSBCs compared with consultant units (Walsh and Downe 2004). Qualitative research is in its infancy with only a handful of ethnographic and grounded-theory papers completed to date. There is clearly a need for more research from within both paradigms.

The birth centre studied was situated in the Midlands of England and catered for the births of around 300 women a year. It was sited within a small district hospital with the nearest consultant maternity unit fifteen miles away. Midwives and maternity care assistants (MCAs) who staffed the facility provided a twenty-four hour service. There were three birth rooms and capacity for five postnatal women. After gaining National Health Service Research Ethics Committee approval for the study, participant observation was undertaken over a nine-month period and included all hours of the day and all days of the week. In addition, an opportunistic sample of thirty women was interviewed approximately three months after giving birth at the centre. A purposive sample of ten midwives and five MCAs, representing a breadth of clinical experience, was also interviewed.

All interviews and field notes were recorded and transcribed. Analysis in ethnography happens concurrently with data collection. This dialectical interaction between data collection and analysis requires the researcher to carefully consider both activities simultaneously while maintaining an open orientation to new data. Thematic distillation was arrived at through this process.

## Fordism and Taylorism in Maternity Services

The very first observational session at the birth centre alerted me to a qualitative difference in how care was organized and undertaken there compared with larger maternity units, which had been the basis of my experience up to that point. A field note captured this moment:

> I'm having a mini crisis here. I've had four cups of coffee and been offered more and it is not even lunchtime. There are no women in labour and just three postnatals. Either this place is over resourced or under used or something else but I don't know what that is yet.

Coming from a practice background of a busy consultant maternity unit, I was steeped in a culture of busy-ness and had to confront on day one of data collection something entirely different. At the end of that day a new perspective was struggling to emerge as a further diary entry recorded:

> I know about 'process mentality' in maternity hospitals and I am very critical of it, so why does it feel so strange to be in a place where processing is not in the vocabulary. I can see already that the quality of the interactions among the staff, and between the staff and the women, is different. Should they just increase the throughput a bit so that there is more stuff to do?

Upon reflection, it was very clear that large maternity units, where anywhere between 4,000 and 8,000 births occurred each year, structured their work according to stages so that women could be moved through at an acceptable rate. The primary rationale for this sequential structure that had evolved over time was said to be clinical. Labour must be completed within a set number of hours, otherwise outcomes were poorer for mothers and babies. Freidman (1954) had established benchmarks for acceptable progress in labour in the 1950s and these have been refined since (see Chapters 3 and 4, this volume). Later, O'Driscoll and Meager (1986), working in Ireland, utilized all these contributions in establishing a labour management protocol that set stringent upper time limits for the length of labour and endorsed early and aggressive measures to speed up labour if it slowed down.

Though some childbirth activists railed against the assembly-line motif that these labour management protocols resulted in (Martin 1987), it was not until Perkins' (2004) recent influential critique that it was proposed that an organizational imperative was equally as important as clinical efficacy in driving concerns about the length of time spent in labour. If women were not moved through the system at a particular rate, then the assembly line backed up and broke down. In fact, recent studies have largely discredited Freidman's and O'Driscoll's findings (Albers et al. 1999; Thornton 1996), reinforcing the conclusion that organizational issues form the principal reasons this style of care continues. Parallels with the industrial model of Fordism are striking: both involve breaking down activities into component parts, with each part undertaken separately by different employees who have clearly demarcated roles. As a car is 'birthed' following linear and discrete processes on an assembly line, so labouring women are processed through 'stages', using a mechanistic model.

Women interviewed in the birth-centre study echoed this kind of language. Agnes was commenting on her perception of birth in a nearby consultant unit:

> At the consultant unit, you felt almost like you were on a conveyer belt and all the nurses were a bit robotic towards you.

Deirdre, a midwife, used the same metaphor in her interview:

> Consultant units are like baby factories.

Procrastination, delay, tangential activities and idiosyncratic patterns cannot be accommodated because of the knock-on effects for other stages of the labour. Hunt and Symonds (1995) observed, in their study of a delivery suite in a large consultant unit, that the labour 'procrastinators' ('nigglers' or women in early labour) did not constitute real work in the eyes of midwives in their study, and that this activity needed sifting out if the system was to work efficiently. Delays after a process was started were dealt with by acceleratory interventions like the artificial rupture of membranes.

Labour care in the birth centre responded to the 'nigglers' in innovative ways that did not posit them as a 'problem', as the following story reveals. A woman, expecting her first baby, came in at midday. Her husband was with her but she was really in the very early stages of labour so they decided to return back home. They came back in the early evening but again her cervix was just two to three centimetres dilated. She was contracting but comfortable. Her husband had a commitment as a DJ for a local rugby club that evening and they decided he should honour it. The woman stayed behind until about 9 P. M. but was becoming bored and 'fed up'. The field note entry continues:

> She says 'I think I would just rather go and be with him', so she went and sat with him at the rugby-club do. He's doing the DJ-ing and she is at the

back, sitting down and while all that's going on she is obviously quietly
labouring because when she comes back at 12.30 A.M., she delivers.

Here, the woman in early labour was not sent back to her home to await
the moment when she required serious midwifery support, but expressed
her own agency in choosing a totally different environment – a night club.
In fact, she experienced the majority of her labour there, arriving back at
the birth centre for the labour's conclusion, the birth itself. Although this
is not typical of what many women would want, nor what midwives
expect, her choice of how to spend the time in labour was respected.

There were a number of examples of tangential activities at the birth
centre that would probably not be provided, or which would be ignored
or cause irritation, in many large hospitals. Many of these were to do with
the needs of the woman's family or friends. Birth-centre staff looking
after the children while their mother was labouring was one example.
Preparing food for birth partners and caring for family and friends who
felt faint or sick were others.

If Fordism represented an industrial model of structuring care in
maternity hospitals, then Taylorism emerged as the scientific management
approach to overseeing its implementation (Dubois et al. 2001). Taylorist
management was hierarchical, detached from worker activity, had a
strong regulatory function, and was focused on product outcomes. Its
values were predictability, standardization and efficiency. In Taylorist
organizations, role differentiation was explicit and tasks were procedure-
driven. Again, Hunt and Symonds (1995) identified these characteristics
in their delivery-suite ethnography. Efficiency on a labour ward was
achieved when procedures were adhered to, hierarchical authority
was respected, and role compliance achieved. The result was, from the
organizational viewpoint, women flowing seamlessly through the labour
and birth process. In addition, Taylorism introduced the idea of time and
motion studies as an indicator of worker efficiency. Tietze and Musson
(2002) argue that Taylorism's unrelenting focus on the external monitoring
of time effectively commodified it, giving us the phrase 'time is money'.
From this perspective, employee efficiency could be measured according
to task completion within set time bands. Time has therefore evolved to
become a critical barometer for both labour management and employee
productivity. Before critiquing these developments in labour and birth
care, and exploring a different perspective in the birth-centre context,
I will outline popular notions of time as understood in contemporary
culture and some of their application in the healthcare context.

## Theories of Time

Dunmire (2000), drawing on Davies's (1990) conception of temporal
models, wrote of two time orientations, cyclical and linear. The first is
historically linked with agrarian society. Labour is determined by the

number of daylight hours. Subsistence pursuits are regulated by the seasons and the reproductive cycle. Events and activities are paced according to a local or natural rhythm with the task determining the amount of time required. In cyclical time, the answer to the question: 'how many hours/days will this take?' is 'as long as it takes to complete it'. This is 'process time' and contrasts sharply with the concept of linear time that proliferated during the rise of industrial capitalism and which construed time as an unfolding straight line, unidirectional and a rational regulator of tasks. This is 'clock time' as we understand it now, setting the parameters of activities, especially work and social life, coming before the task or event, not as a consequence of these. It is a decontextualized temporality and allows the individual to 'make time' and have 'free time': the first by completing the task before its expected time allocation, the second by dichotomizing life into work and leisure.

What is deceptive about linear time is, although it appears to operate outside culture, beyond social and national boundaries, its realization is laden with contextual provisos. The capacity to 'make time' as an assembly-line worker may simply mean higher productivity for industry with more being done in the same allocated time. A manager, by contrast, who 'makes time' by completing an activity early is more likely to have discretionary time which they can spend as they choose. In addition, in relation to the work/home dichotomy, men are more likely than women to have free time away from paid employment because of the gendered nature of the home environment (Graham 1983).

Gender differences in the experience of 'lived time' has prompted Lyon and Woodward to suggest that linear and clock time is masculine, 'easily broken into its component units, of single usage, dimension and direction, fitting well with the rational working practices of the Weberian bureaucracy' (2004: 209). They aligned 'process time' with the feminine, characterizing it as relational, continuous and cyclical. This dichotomy opens them to the charge of essentialism were it not for the grounding of their views in an established feminist critique of the public/private spheres (Kimmel 2000). This critique has exposed these gendered spheres as privileging men in the public and discriminating against women in the private (Graham 1983). Everingham (2002) extended the critique to considerations of temporality in these respective spheres, suggesting that in the private sphere, domestic duties 'flow' with no obvious end point and are characterized by repetitive rhythms. Mothering, undertaken primarily in the domestic sphere, is biologically time-determined, both in its gestation and enactment, responsive to the rhythms of an infant who is not yet accustomed to clock time. Men's time as a traditional breadwinner in industrialized societies, is linear. Labour, sold by a temporal denominator (hour, day, week, month, year) commences and ends at specific times. Once completed, time becomes 'time off' or 'free time' to be filled as a reward for employment time.

Everingham (2002) was primarily concerned with gender equity in the workplace and problematized contemporary notions of flexibility and family-friendly policies which are trumpeted as solutions to gender inequality. I refer to her work here because of its relevance to the following discussion of temporality in the birth centre and maternity services more broadly, a setting where it could be assumed biological and cyclical time are clearly in the ascendancy, but where, as already mentioned, an organizational imperative, in addition to clinical factors, drive a preoccupation with linear time.

There is little written about temporality in the healthcare context and, where it is discussed, assumptions have been made about the centrality of linear time in structuring caring. Waterworth (2003) is a notable exception and much of what she discussed resonates with maternity care. Critiquing the reductionist approach of linear models, her research engaged the differing temporal reference frameworks of the individual nurse and of nursing teams working in hospital settings. She found tension and conflict between these frameworks as individual nurses attempted to balance relational time spent with patients against clock time pressures of 'getting through the work' for the sake of the team. Though time management was viewed as an individual nurse's responsibility, it was inscribed with social imperatives concerning the wider milieu. When tasks are left undone because a nurse has traded relational time spent with individual patients with time for routine tasks, social sanctioning from the team ensues. Similar observations were made by McCourt (2006) in an observational study of midwives' communication in the clinic. The only acceptable reasons for 'slippage' were emergencies and marked increase in busy-ness related to patient throughput. The dominance of linear time over process time was also demonstrated in performance measures of nursing effectiveness: their skills in prioritising, their allegiance to team goals and their ability to 'catch-up'. Even in palliative care settings, where the legitimacy of patients needing time and where listening to patients was enshrined in the nursing care philosophy, relational time to care for emotional needs was subordinate to work (task) demands (James 1989).

Ball et al.'s (2002) exploration of why midwives leave the profession in the U.K. painted a similar picture of midwives having to 'get through the work', of having to rush to 'catch up' and of the guilt felt at leaving work for colleagues on later shifts. The unwritten objective of getting all the 'checks' done on the early shift was found in the studies of both Ball et al. (2002) and Waterworth (2003).                                               .

Other layers of social meaning impinge on linear time as experienced by nurses and midwives in practice. Rhetoric from health policy around patient-centred care, continuity of care and individualized care all require more 'process time' but must be subjected to linear time. Thus, as Fitzpatrick (2004) argues, there exists a hierarchy of time domains. 'Absolute time' (linear time) is not morally neutral and obscures the power relations inherent in temporality. This theme will be developed in the next section, specifically examining the findings from the ethnographic study of the birth centre.

## Temporality and Care in the Birth Centre

### *Labour and Birth*

Birth-centre care was typified by an absence of task orientation and time-defined routines. In relation to labour care, this manifested itself less as a concern for the progress of labour and more as a focus on environmental ambience and the women's embodied experience of labour.

The centrality of the environment to the staff's priorities was startling. The building had been honed over a ten-year period so that a traditional maternity hospital had been transformed into a 'homely' birth centre. Every room had been 'madeover' to reflect a warm, homely and informal setting for birthing. All of the fund raising and much of the actual decorating had been done by the staff themselves. Their interviews suggested what the effects of the changed environment on the birthing women were. Margaret, a maternity care assistant, told me:

> The women often say 'I've been to other hospitals and they're nothing like this'. 'This really feels like the place I would like to have my baby in'. I have even had mums talking to their babies, saying 'This is where we want to be, in here?'

Louise, a midwife, explained it like this:

> In relation to women feeling intuitively that it is right to give birth here, I think it is similar to animals finding the right spot to give birth and I just think – yes, it feels right, then they will do it.

Women's comments confirmed these impressions. Jasmine, about to have her first baby, said:

> We went to have a look, and as soon as we walked in we thought - yep! This is the sort of place ... because I think it was so small and it's not like a hospital, we thought it would be a nice relaxing place to go.

Another woman said 'I got stuck on it' and 'I could picture myself there'. This idea was developed by others, such as Theresa, who was having her first baby:

> The psychological effects of being there, it was like being at home really in terms of environment, it was very, very comfortable and calming, relaxing. The room itself, the way it's made up. It's got homely things in it. Most of the instruments are hidden away, so that helped. And it's just the atmosphere. It's really something you can't definitely put your finger on, which makes it so difficult when you're talking about it because it's so much in a woman's own mind. It's a feeling rather than an empirical value system. A woman knows immediately when it's the right atmosphere.

In searching for the origin of the staff's drive to optimize the surroundings for birthing, and after hearing the responses of women upon visiting, I concluded that it may come from a 'nesting instinct'. In older midwifery textbooks, nesting is ascribed to the period of early pregnancy when the body's physiology inclines the woman to reduce activity because of fatigue and morning sickness (Myles 1981). As the baby grows, consideration of place of birth becomes more central to women's concerns. It is during this time that the visit to the birth centre usually occurs (between eighteen and twenty-eight weeks).

The idea of the centre 'being like home' was expressed by other women. 'It's like walking around your own house' or 'like being in your own bedroom'. 'Treat the place as your own the midwife said to me', recalled another woman. The home motif may explain the apparently intuitive, nesting-related responses of women when viewing the centre for the first time.

Older textbooks referred to nesting behaviours again in the weeks leading up to the start of labour when many women have finished public employment and are active around the home, preparing for the arrival of the baby. This preparation mirrors the intent of the birth-centre staff. In this sense, they are always preparing for a birth with the ebb and flow of labouring women. Attention to the environment is preparatory work, an ongoing dynamic of activity at the centre. It resembles a rhythm, responsive to the biological cycle of parturition, rather than the compartmentalized routines of an institution. Sometimes it is a flurry of activity, such as when a national television programme was invited in to makeover a room. It was converted to a water-themed birth room over four days. Or when the lead clinical midwife decided to upgrade a room to provide for complementary therapies:

> I just thought, well, why don't we have a Complementary Therapies room then? This room is not really used for anything because it is so horrible you wouldn't want to deliver in here, and then before we knew it, as it always happens here, somebody comes up with an idea and six hours later we have got a master plan. I got the money from petty cash. I went out and did the shopping. We just put up the paper and painted it. It took us three days.

This last story is almost subversive because the 'task' was completed much more quickly than the bureaucratic processes in hospitals, which are predicated on efficiency, would allow. Cyclical temporality dominates here as a project is completed rapidly because the staff want the change to happen. They work on the project over a twenty-four hour period, allowing for care demands, and come in on their days off to complete it. Their work is not controlled by linear parameters of daytime hours and weekends off that impinge on external decorators. On other occasions, change occurs more leisurely, over time, as when a sister of one of the Maternity Care Assistants (MCAs) offered to replace all the curtains in the centre. It took nine months.

The similarities to Everingham's (2002) domestic rhythms are striking in relation to two other dimensions of environmental nurture, hospitality and upkeep. Welcome and hospitality involved some planned, but many unplanned visits by a variety of people – colleagues, external visits, women clients, maintenance people, friends and family. Visitors were always welcomed warmly and offered a drink, and often biscuits, cake or toast. In a similar way, the attention to environment regarding cleanliness, tidiness, decor and ambience was an accepted priority. Many of the visitors were maintenance people attending to a variety of building problems. The communication book was just as full with building-related issues as it was clinical issues, and these were handed on to the staff coming on the next shift in exactly the same way that outstanding clinical issues were. Yet hospitality and upkeep were seldom mentioned in the staff interviews. It just happened as part of their lived experience at the centre and was not articulated as tasks to be done by certain individuals. Everybody contributed, with maternity care assistants (MCAs) and midwives being completely interchangeable. Midwives vacuumed floors, washed curtains, and made drinks. MCAs answered phones, conducted tours and sorted out the engineering problems. As Everingham (2002) comments, these domestic-style activities are open-ended and are not time led as such. They are recurrent like the ebb and flow of a small community and arise from communitarian values and a focus on relationships.

Similarly, labour support was conspicuous in its interpersonal rather than clinical dimension. The priority given to individual women's agency rather than conventional clinical management was graphically illustrated by the following episode, related by a midwife:

> It was the week before Christmas and I had one lady who was five centimetres when she came in. Actually she was really more six [sic] but she was desperate to get her Christmas shopping done – you know she had this little window of time to do it and now this! So because the labour wasn't that strong we decided she could go shopping and come back afterwards. She came back and delivered a couple of hours later.... I was still here when she came back and she got [sic] her shopping done and then she went home that night after the baby was born. You have got to be flexible here.

It is highly unlikely that a woman in established labour in a consultant unit would be 'allowed' to go shopping because linear time predicts that she may have the baby within the next couple of hours. Here, the woman's own reading of her labour rhythm was trusted and the clinical imperative was subjugated to that.

Another labour episode reveals a poignant example of relational temporality. A teenage girl came in with her mother and sister. The girl became extremely distressed in the middle of her labour and the midwife who was assisting with the birth recounted the following:

> She was thrashing around on the bed so we took the bed out. Bev [the midwife] wondered whether her distress was due to the awesome responsibility of parenthood that she felt she wasn't ready for, so Bev asked her mum and her sister to leave the room. Then she [Bev] just sat with her for two hours on the floor and this girl was just sobbing into her lap, just sobbing, and then after two hours – almost as if it was out of her system – she was completely more focused and she went on and had a really good birth.

Again it is hard to imagine this scenario playing out in a large labour ward where distress in labour is usually assumed to have a clinical cause or require pharmacological pain relief. Though relational temporality and labour care is affirmed at the macro level by the accepted efficacy of one-to-one support (Hodnett et al. 2007), it is frequently denied at the micro level of care delivery because of the failure to address staff shortages (Walsh 2004) and the dominance of the 'biomedical metaphor' (Machin and Scamell 1997).

### From 'doing' to 'being'

The absence of clock-time pressure and an assembly-line mentality manifested itself in an unhurried approach to care within the birth centre. Women's frequent use of words like 'relaxed', 'laid-back', 'informal' points to the almost tranquil ambience that they perceived. A dismantling of an assembly-line approach was also demonstrated by a deregulated approach to the common routines of consultant units, like morning mother and baby checks, regular observations and set procedures around meals and bathing. There were no bells to herald the end of visiting times and women never used buzzers to summon staff attendance. No drug rounds were done because women self medicate. In the mornings, they basically got up whenever they wanted to. Some would make it to breakfast, served buffet-style from 8 o'clock onwards. The staff told me that the teenage mums liked a lie-in, sometimes until early afternoon if their babies did not wake them. Baths and showers happened at the woman's convenience. The jacuzzi baths were very popular, lasting up to an hour at a time with some women having two per day.

Though daily 'checking' of mothers and babies was done, it was conducted in the context of a 'chat'. In fact 'chatting' was what much of the communication between staff and women consisted of, as indicated by this field note entry:

> The midwife actually spent a fair bit of time chatting and talking.... I kept a log of her activities roughly on and off through the day so that I could get a sense of what sort of routine there was. Very deregulated. The way they did the postnatal check was fairly traditional in one sense, but not comprehensive at all in the sense of stripping the baby down, or, you know, doing temperature, or doing top to tail, head to foot check on the women.

A later entry records:

> Conscious of how the staff here can chat with the women in a very friendly way and have time to laugh and have a joke and talk to the other siblings and engage the husband and chit-chat. There is a fair bit of just chit-chat that goes on generally as they go and see the women and see how they are

The log of activities referred to in these observations recorded 'talking and chatting' to various people repeatedly throughout the day – women, their relatives, staff, visiting community staff, maintenance men, external visitors, clerical staff. Women liked the chatting as this transcript indicates:

> Belinda [MCA] came and sat with a couple of us. She has had about five children herself. She just chatted and talked. Because when the baby is so new and everything, you are sort of quite confused with all talk of babies. She just was chatting to us and she was just really lovely.

This kind of interaction occurred sometimes in the context of a task, like making a bed, helping with a baby bath, but sometimes while just hanging around the women. On night duty it seemed to happen more often, because women were frequently awake tending to their babies, and the night staff would wander up to their room to enquire if they were all right. One of the women recalled:

> At 10 o'clock at night we'd put the babies down and get round the telly and have lots of, like, cocoa, and it was so relaxed. And, like, the midwife came down and joined us and often bought us drinks and some Kenco biscuits. She was, like, chit-chatting, just being a friend.

This informal 'talk' seemed also to be aligned with allowing women to care for their babies and adjust to the post-birth period in their own time and at their own pace. A number of women mentioned that there was 'no nagging', 'no fussing' and that they were trusted to 'get on with it'. If the staff did intervene, then they would demonstrate something and then 'step away'. When staff came down to see the women, it wasn't for checking up on them but to ask them how they felt.

Talking or chatting that is not task-related or problem-centred, but something that is done naturally while just sharing the same space as the women, can be understood as a dimension of 'being' rather than 'doing' (Fahy 1998). It captures something of the essence of 'with woman' that midwifery is predicated on. Fahy argues that the emphasis on 'doing' is a product of the modernist, techno-rational scientific approach. 'Doing' can also be seen as a key behaviour in Fordist approaches to production. 'Doing' is the institutional work based on set routines and tasks portrayed in many studies of maternity units (Bick 2000; Kitzinger, Green and Coupland 1990). 'Being', though, is more clearly aligned with a relational disposition that values the emotional and the subjective. One gets a sense of this focus from the open-ended and, in a sense, 'purposeless' interactions described here.[1]

Hanging around with the women ('waiting on' them) could be viewed as unnecessary or redundant activity if judged according to traditional performance indicators of time-and-motion analysis of staff activity. These measures are inevitably skewed towards case-mix and dependency analysis so that clinical areas with a higher percentage of women with complications are judged to require more staff. This contributes to a perception that midwifery-led units and birth centres will require fewer staff. Even the well-validated tool, Birthrate Plus (Ball and Washbrook 1996), has struggled to assess appropriate staffing levels for small midwifery-led units (Ball et al. 2003). But 'being with' is obviously a time rich activity, as the above quotes from women and the story of the teenage girl who was held in an extended embrace indicate . It clearly connects with 'process time', the affirming of the relational and intersubjective.

This kind of 'being' resonates with a related theme that Leap (2000) has written about in her paradoxical explication of supportive midwifery care, specifically in the childbirth context, of 'doing more by doing less'. Kennedy (2000) echoed the same idea in a Delphi study of expert midwives who were asked to articulate core skills for midwifery care. 'Doing nothing well' was a recurrent theme from her data. Leap (2000) and Hunter (2004) both adopt the active verb 'doing', central to the Fordist and Taylorist model and integral to any application of the linear time paradigm, but subvert its usual meaning to highlight the benefit of 'being with'. 'Being with' resists measurement and monitoring yet hints at important interpersonal values like altruism, respect and solidarity. 'Being with' as a disposition for a person providing one-to-one care in labour connects powerfully with Chalmers' (1989) wise statement, made upon the initiation of the Cochrane database of evidence for childbirth and paraphrased here, 'what really counts in childbirth care cannot be counted'. 'Being with' is constituent of a fluid temporality and, in the childbirth context, dependent on biological rhythms and interpersonal dynamics.

## Gender and Power in Birth-centre Temporality

Temporality has been described by Everingham (2002) as a deeply gendered concept and by Fitzpatrick (2004) as suffused with power differentials. Both refer to the hegemonic properties of linear time which seeks to colonize all human activities while at the same time masquerading as value neutral and objective. Everingham reveals the dishonesty of this and the social patterning of time that segments public and private, work and non-work, and falsely implies that time spent can be reduced to measurable units. Feminist critiques of these ideas centre around 'home work' being multi-layered (emotional labour alongside the physical labour of childcare) and polychronic (the simultaneous doing of several tasks, or multi-tasking) which are both resistant to disassembling (Lyon and Woodward 2004). Organizational critiques have examined the deregulation of workplace hours and the adoption of family friendly policies, both of which purport to give employees more flexibility to balance their work and home lives (Warren 2004).

The birth centre data addressed these areas and the application appeared to be empowering rather than disempowering to staff, increasing their agency in these areas. I got a sense of this very early on in staff interviews, as the following comment from Gerry suggests:

> Working here is like having your favourite chocolate bar. You really want it, you get to have it and you still want some more. It's lovely.

These sentiments were to be repeated with slightly different emphases many times as the interviews continued. 'I thank God everyday that I can work here', said one midwife. 'I waited for ten years for this - it's my dream job' commented one of the MCAs. Another midwife likened working there to 'practising perfect midwifery', while another offered 'I just love working here, it's like a breath of fresh air'.

As already indicated, language around 'home' was common in women's descriptions of their experiences at the birth centre, and staff activities, conceptualized as nesting and environmental nurturance, were also suggestive of a domestic domain. In addition, the pace of care, likened to a 'being' rather than 'doing' disposition, resonates with the cyclical temporality of home life. Evidence for a 'lived' notion of time also comes from examples of a blurring between work and home which deconstruct the dichotomous work/life split, derivative of linear temporality. I recorded the following field note:

> One midwife's husband and two daughters came in. He couldn't manage to do their hair before taking them out horse riding so the midwife said she would. The two daughters sat in the office, made themselves at home while she French-plaited their hair.

On another occasion, one of the MCAs told me that the very young children of one of the midwives found it hard to settle at night if she was on a late shift. Her husband used to bring them into the unit around seven and she would go through a bed-time routine with them before he would return them home.

There were at least three other occasions when I observed family members coming in for a chat or on an errand while I was there. It appeared to be quite a normal event in the activity of the birth centre. In addition, staff would come in on their days off for a coffee if they were in town shopping and there was clearly an explicit social, community dimension to the centre. It was a meeting place. All of these activities transgress traditional boundaries between working and not working, between being 'on duty' and 'off duty'. In fact, some of the activities were specifically home/family mediated.

The most radical challenge to the work/life split was an experience related to me by one of the MCAs who mainly worked nights:

> The girls here have been marvellous to me, I have to say. Problems that I have had at home with elderly relatives dying of cancer. I've had a load of hassle and they have been really good. I've come in on nights and been knackered, not had any sleep and they have tucked me up in bed for a couple of hours.

Sleeping while on duty could be a sackable offence was how some midwives responded to this story when it was related to them. Yet, in its context, it appeared a compassionate act that subjugated Taylorist time regulation to a relational imperative.

The staff spoke of the blurring of work/life boundaries with paradoxical language:

> When you're here, it becomes your life. When I started, I felt like I was coming home. When I come to work, it is not like going to work, it's like going home.

Such language is suggestive of 'process' time where life activity flows and integrates, if not resisting clock time inscription then certainly challenging its regulatory function.

In addressing Everingham's (2002) concern that emotional work and multi-tasking are rendered invisible in the private sphere, activities at the birth centre explicitly validate their worth, as a number of the stories retold here indicate. A midwife, new to the birth centre, expressed her surprise at the variety of organizational tasks as the following anecdote indicates:

> The plumbers arrived with the birthing pool and asked me where they should put it. I thought 'do I have the authority to make that decision?' I rang one of the other midwives and she said, 'Yes, you do – you're on duty, it is your call'. I'm not used to that level of authority. But then I realized, I make those kinds of decisions at home all the time.

Her words are just one example of a plethora of metaphorical images aligning the birth centre to 'home' and the interesting crossover of skills and the level of autonomy between both settings which the midwife here finds empowering. Some feminist critiques of the gendered nature of the private and domestic spheres speak of the impact that this has had on women working in paid employment. Hessing (1994) referred to the 'double shift' as women are often burdened with two full-time roles. She also wrote of how the autonomy of home is not usually matched by women's status in the workplace, where they tend to get lower-level jobs with little responsibility. At the birth centre, the staff have real autonomy in day-to-day decision making and control over their working conditions. In particular, the effects of the 'double shift' are mitigated here by the part-time status of most of the workforce. The balancing of responsibilities at both work and home has been structurally addressed by employer initiatives collectively known as family-friendly policies (Scheibl and Dex 1998), which include discretionary provision for part-time hours. However, it is the scale of the provision that stands out at the birth centre where only two members of staff are full-time.

Family-friendly policies, writes Everingham (2002), are of less significance than measures that can be adopted by employees to give them greater control over when and how they work their hours – control over starting and finishing times, influence over the pace of work, access to a phone at work to use for family reasons. I was struck at precisely this level of freedom when I observed work patterns at the birth centre. The informal, day-to-day accommodating of staff members requesting to come in later or go off earlier because of a variety of family, school or other commitments was ubiquitous and the willingness of staff to stay at important times, such as waiting on a birth. When I questioned the staff about this they accepted it as 'the swings and roundabouts' of flexible working. The level of flexibility extended to babysitting networks among staff and mutual childcare arrangements during the weeks of school holidays.

All of these features of work patterns at the birth centre challenge the power of linear, clock-time temporality and place alongside it a balancing force from a cyclical, relational, 'lived' temporality. Though using the regulatory framework of clock time, for example, in relation to staff shift times or the recording of labour progress, time's boundaries in the birth centre are permeable and malleable, often overridden by social, psychological or biologically mediated considerations.

Feminist critiques rightly highlight the gendered nature of the public/ private spheres and problematize family-friendly policies as a response to this. However, the birth centre work appears to express an enabling and empowering integration of the public and private domains and a positive application of family-friendly rhetoric. This is based in part on the altered temporality of these spaces and the way that staff agency and autonomy can be expressed there.

## Conclusion: Toward 'Slow Birth'

Earlier in his chapter, I considered my first reaction to observing a different temporality in the birth centre when compared to my past experience in large maternity hospitals. I suspected that there was something qualitatively different about this environment once the pressure to process women was removed. Now I am closer to articulating that difference and Parkins' (2004) and Honore's (2004) concept of slow living seems particularly relevant. Each was trying to articulate the value of 'having time', viewed from within the dominant linear model, in an attempt to redeem clock time. Using Reisch's phrase 'enough time for meaningful things' (2001:374), Parkins (2004) emphasizes investing time with significance through attention and deliberation. This is not free time to fritter as a passive receiver but a creative and dynamic engagement that enriches oneself and others. To make her point, Parkins (2004) uses the analogy of the 'slow food' movement, which sees itself as an antidote to the 'MacDonaldization' of cuisine and the exponential growth of 'fast food'. The 'slow food' movement emphasizes organic ingredients, traditional cooking techniques and indigenous recipes. It views cooking as an artistic endeavour not to be hurried, with its fruits to be savoured and lingered over.

Slow living of this kind fits the rhythm and flow of labour and birth. To savour, appreciate and ingest the profundity of meaning around childbirth, then it is surely process time, not linear time, which needs invoking. Time-line and volume-driven childbirth is maternity care's equivalent of 'fast food'. It is 'fast birth' and the MacDonaldization of parturition. Birth centre care shows another way, reliant on physiology not technology, using more traditional/intuitive skills and birthing locally, close to kinship and community.

Adam seemed to capture the spirit of my birth-centre study findings on time when she wrote, 'women's time of caring, loving and educating, of household management and maintenance, female times of menstruation, pregnancy, birth and lactation are not so much time measured, spent, allocated and controlled as time lived, time made and time generated' (1995: 95). Thus the birth-centre study gives a tantalizing glimpse of an alternative birth culture within a modern health service setting and, by implication, represents a stringent critique of assembly-line birth with its Taylorist control mechanisms and obsession with the labour time line. By way of contrast, birth-centre care in labour reflects the diversity of women's physiology, psychology and social milieus. These manifest themselves as women are 'waited on' by the staff, who have an attitude and disposition more akin to 'being with' than 'doing to'. The language of home and the experience of community pulse through the accounts of women and staff of life at the birth centre, where temporal space is made for 'slow birth'. Though the birth centre is predominantly a female environment, it does not appear to be gendered in an oppressive way. Women users and staff harness socially defined roles of mothering and nurturing and express creativity, autonomy and agency in fleshing them out. There is a harmony about the birth centre environment which I suggest has much to do with the subjugation of linear time to process time in a reversal of the usual power differentials. Current large-scale maternity services, not only in industrialized countries but also in countries where large scale hospital units modelled on industrial settings have been rapidly introduced and prioritized, have much to learn from it.

## Notes

1.  Models of efficiency may be predicated on the assumption that 'being' cannot equate to 'doing', which seems to be reflected in much of the literature focusing on busy-ness in nursing and medicine (see, e.g., McCourt-Perring 1993). However, an observational study of interaction between midwives and women (see also Chapter 5, this volume) showed that caseload midwives who appeared to be simply chatting to women in a casual and disorganized way in fact scored very highly on task-focused audits. That is, they were achieving a great deal of clinical as well as relational work (McCourt, C. 2006).

## References

Adam B. 1995. *Timewatch: The Social Analysis of Time.* Cambridge: Polity.
Albers, L. 1999. 'The Duration of Labour in Healthy Women', *Journal of Perinatology* 19(2): 114–9.
Ball, L., P. Curtis and M. Kirkham. 2002. *Why do midwives leave?* London: Royal College of Midwives.

Ball, J., B. Bennett, M. Washbrook and F. Webster. 2003. 'Birthrate Plus Programme', *British Journal of Midwifery* 11(6): 357–65. Ball, J. and M. Washbrook. 1996. *Birthrate Plus*. Oxford: Books for Midwives.

Bick, D. 2000. 'Organisation of Postnatal Care and Related Issues', in J. Alexander, S. Roche and V. Levy (eds), *Midwifery Practice: Core Topics 3*. London: MacMillan, pp. 129–42.

Chalmers, I. et al. 1989. *Guide to Effective Care in Pregnancy and Childbirth*. Oxford: Oxford University Press.

Davies, K. 1990 *Women, Time and the Weaving of the Strands of Everyday Life*. Aldershot: Avebury.

Davis-Floyd, R. and E. Davis. 1997. 'Intuition as Authoritative Knowledge in Midwifery and Homebirth', in R. Davis-Floyd and C. Sargent (eds), *Childbirth and Authoritative Knowledge*, Berkeley: University of California Press, pp. 315–49.

Downe S., C. McCormick and B. Beech. 2001. 'Labour Interventions Associated with Normal Birth', *British Journal of Midwifery* 9(10): 602–6.

Downe, S. and C. McCourt. 2008. 'From Being to Becoming: Reconstructing Childbirth Knowledges', in S. Downe (ed.) *Normal Childbirth: Evidence and Debate*, 2nd edn. London: Churchill Livingstone, pp. 3–24.

Dubois, P. , M. Heidenreich, M. La Rosa and G. Schmidt. 2001. 'New Technologies and Post-Taylorist Regulation Models: The Introduction and Use of Production Planning Systems in French, Italian and German Enterprises', in W. Littek and T. Charles (eds), *The Division of Labour*. Berlin: De Gruyter, pp. 287–315.

Dunmire, P. 2000. 'Genre as Temporally Situated Social Action', *Written Communication* 17(1): 93–138.

Everingham, C. 2002. 'Engendered Time: Gender Equity and Discourses of Workplace Flexibility', *Time and Society* 11: 335–51.

Fahy, K. 1998. 'Being a Midwife or Doing Midwifery', *Australian Midwives College Journal* 11(2): 11–16.

Fitzpatrick, T. 2004. 'Social Policy and Time', *Time and Society* 13: 197–219.

Fox, N. 1999. *Beyond Health: Postmodernism and Embodiment*. London: Free Association Books.

Friedman, E. 1954. 'The Graphic Analysis of Labour', *American Journal of Obstetrics and Gynaecology* 68: 1568–75.

Graham, H. 1983 'Caring: A Labour of Love', in J. Finch and D. Groves (eds), *A Labour of Love: Women, Work and Caring*. London: Routledge and Kegan Paul, pp. 132–146.

Hessing, M. 1994. 'More than Clockwork: Women's Time Management in their Combined Workloads', *Sociological Perspectives* 37(4): 611–33.

Hodnett, E.D., Gates, S., Hofmeyr, G.J. and C. Sakala. 2007. *Continuous support for women during childbirth*. Cochrane Database of Systematic Reviews 2007, Issue 3. Art. No.: CD003766. DOI: 10.1002/14651858.CD003766.pub2.

Holub, R. 1992. *Antonio Gramsci: Beyond Marxism and Postmodernism*. London: Routledge.

Honore, C. 2004. *In Praise of Slow: How a Worldwide Movement is Challenging the Cult of Speed*. London: Orion.

Hunt, S. and A. Symonds. 1995. *The Social Meaning of Midwifery*. Basingstoke: Macmillan.

Hunter, B. 2004. 'Conflicting ideologies as a source of emotion work in midwifery', *Midwifery*, 20(3):261–72.

James, N. 1989. 'Emotional Labour: Skill and Work in the Social Regulation of Feelings', *Sociological Review* 37: 15–42.

Kennedy, H. 2000. 'A Model of Exemplary Midwifery Practice: Results of a Delphi Study', *Journal of Midwifery and Women's Health* 45(1): 4–19.

Kimmel, M. 2000. *The Gendered Society*. Oxford: Oxford University Press.

Kitzinger, J., Green, J. and V. Coupland. 1990. 'Labour Relations: Midwives and Doctors on the Labour Ward', in J. Garcia, R. Kilpatrick and M. Richards (eds), *The Politics of Maternity Care*. Oxford: Clarendon Press, pp. 149–62.

Leap, N. 2000. 'The Less We Do, the More We Give', in M. Kirkham (ed.) *The Midwife–mother Relationship*. London: Macmillan, pp. 1–18.

Lyon D. and A. Woodward. 2004. 'Gender and Time at the Top: Cultural Constructions of Time in High-level Careers and Homes', *European Journal of Women's Studies* 11(2): 205–21.

McCourt, C. 2006. 'Supporting Choice and Control? Communication and Interaction between Midwives and Women at the Antenatal Booking Visit', Social Science and Medicine 62(6): 1307–18.

McCourt-Perring, C. 1993. *The Experience of Psychiatric Hospital Closure: An Anthropological Study*. Aldershot: Avebury.

Machin, D. and M. Scamell. 1997. 'The Experience of Labour: Using Ethnography to Explore the Irresistible Nature of the Biomedical Metaphor during Labour', *Midwifery* 13: 78–84.

Martin, E. 1987. *The Woman in the Body: A Cultural Analysis of Reproduction*. Milton Keynes: Open University Press.

Mead, M. 2008. 'Midwives' Practices in Eleven UK Maternity Units', in S. Downe (ed.) *Normal Childbirth: Evidence and Debate*. 2nd ed. London: Churchill Livingstone, pp. 81–96.

Myles, M., 1981. *Myles' Textbook for Midwives*. Edinburgh: Churchill Livingstone.

O'Driscoll, K. and D. Meager. 1986. *Active Management of Labour*. London: W.B. Saunders.

Parkins, W. 2004 'Out of Time: Fast Subjects and Slow Living', *Time and Society* 13: 363–82.

Perkins, B. 2004. *The Medical Delivery Business: Health Reform, Childbirth and the Economic Order*. New Brunswick, NJ: Rutgers University Press.

Reisch, L. 2001. 'Time and Wealth: The Role of Time and Temporalities for Sustainable Patterns of Consumption', *Time and Society* 10: 367–85.

Scheibl, F. and F. Dex. 1998. 'Should We Have More Family-friendly Policies?' *European Management Journal* 16(5): 586–99.

Tew, M. 1998. *Safer Childbirth? A Critical History of Maternity Care,* 3rd ed. London: Chapman and Hall. Thornton, J. 1996. 'Active Management of Labour', *British Medical Journal* 103:313– 378.

Tietze, S. an G. Musson. 2002. 'When 'Work' meets 'Home': Temporal Flexibility as Lived Experience', *Time and Society* 11: 315–34.

Wagner, M. 2001. 'Fish Can't See Water: The Need to Humanize Birth', *International Journal of Gynaecology and Obstetrics* 75: S25–S37.

Walsh, D. 2004. 'Home Birth, Staffing and Acute Services', *British Journal of Midwifery* 12(10): 616.

Walsh, D. and S. Downe. 2004. 'Outcomes of Free-standing, Midwifery-led Birth Centres: A Structured Review of the Evidence', *Birth* 31(3): 222–9.

Warren, T. 2004. 'Working Part-time: Achieving a Successful Work–life Balance? *British Journal of Sociology* 55(1): 99–122.

Waterworth, S. 2003. 'Temporal Reference Frameworks and Nursing Work Organisation', *Time and Society* 12(1): 41–54.

*CHAPTER 7*

# MANAGEMENT OF TIME IN ABORIGINAL AND NORTHERN MIDWIFERY SETTINGS

*Gisela Becker*

## Introduction

This chapter explores Aboriginal concepts of time and their impact on maternity care for women and babies in a variety of midwifery settings in Canada. I describe the experiences of time management in a variety of Aboriginal cultures, including the Inuit in Nunavut and Nunavik (northern Québec), the Dene and Métis in the Northwest Territories, and the Six Nations in Ontario. Aboriginal concepts of time differ from non-Aboriginal concepts, where time is mainly seen as a linear concept. The Aboriginal concept of time may also vary, but in general terms situates events in a circular pattern of time, in which the person is at the centre of time circles and events are placed in time according to their significance to the individual and the community. This view of time impacts on women's, families' and midwives' experiences of childbirth when providing culturally safe maternity care. Women living in remote settings and Aboriginal women also face other challenges, such as socio-cultural and geographic influences that affect their lives and health.

## The Population and Health of the Aboriginal Community in Canada

Of the various cultural groups in Canada, the Aboriginal community is a rapidly growing one. Data from the 2001 census shows that the number of Aboriginal people as a percentage of Canada's total population is on the rise. Today, the Aboriginal population is younger than the non-Aboriginal population, suggesting that fertility rates are higher. However, population increase alone cannot account for the growth, some of which is a result of the change in the way Aboriginal people have responded to recent census questions about Aboriginal origins or identity (Siggner 2003). Over 1.3 million people reported having at least some Aboriginal ancestry in 2001, representing 4.4 per cent of the total population, while in 1996 people with Aboriginal ancestry represented 3.8 per cent of the total population (Statistics Canada 2003).

The highest concentrations of the Aboriginal population in 2001 were in the north and on the prairies. The 22,720 Aboriginal people in Nunavut represented 85 per cent of the territory's total population, the highest concentration in the country. Aboriginal people represented 51 per cent of the population of the Northwest Territories, and 23 per cent of the population of the Yukon (Statistics Canada 2003).

Presently, the Aboriginal community in Canada faces serious health issues, including high suicide rates, epidemic proportions of diabetes and high levels of alcohol and drug use.[1] In traditional Aboriginal cultures, women are considered to be the givers of life and traditionally this role in the family was highly respected.[2] (Grace 2002). However, many Aboriginal women today face a higher level of risk of multifaceted health problems than women in the general population (Statistics Canada 1998).

Rapid culture changes within the Aboriginal and non-Aboriginal communities and a marginalized position in today's society have contributed to the increased health problems of Aboriginal women (INA 2000). Aboriginal girls and women face shorter life expectancy and higher rates of suicide, diabetes, cervical cancer, sexually transmitted infections, mental health problems and other indicators of morbidity when compared with non-Aboriginal women in Canada.

The roots of such health problems have been identified by a number of writers as lying in the history of colonial and postcolonial contact with non-native peoples (Kelm 1998). By the early twentieth century, dramatic population decline had occurred as a result of contact with infectious diseases introduced by Europeans. Later in the twentieth century, this was followed by rises in chronic health problems that had previously been less common in Aboriginal communities associated with changes in lifestyle, poor living conditions, and poor mental health experienced by Aboriginal communities (Waldram et al. 2006). Such health problems were compounded by the socio-economic, cultural, physical and psychological impact of colonization, including enforced settlement and the removal of children to residential schools (Kelm 1998). Kelm (1998) noted that the 'Aboriginal body' came to be seen by settler communities and government officials as unhealthy and in need of management or reform. This view extended to women's health and their conduct as mothers. Consequently, the anthropologist O'Neil stated in 1988 that the health services are 'one of the most powerful symbols of the colonial relationship between Northern peoples and the nation state' (cited in Lavoie 2001: 332). Kelm argued that the residential schools were consciously designed to transform Aboriginal bodies and culture, to draw children away from their home influences. 'In the eyes of their advocates', Kelm argued, 'residential schools, and the acculturative training they offered, would save a "race" dying from maternal neglect' (1988: 61). However, the massive losses of life, and of family and culture, led to multiple health problems and a need for 'community healing' (Degnen 2001).

## Aboriginal Concepts of Time and Space

The perception of time may vary in different Aboriginal communities since it cannot be removed from the cultural context of each distinct people. However, some concepts, including certain understandings of time, are common to the Canadian aboriginal community. The predominant Aboriginal concept of time is intuitive, personal and flexible and many Aboriginal people do not seem to mind if social events or other meetings start later than scheduled (Brant 1990). Having lived rather dependently with nature in the past, people employ the concept of 'doing things when the time is right', when all environmental factors combine to bring success. Today, the concept of time is less a principle of living and more an expression of the need to maintain harmonious interpersonal relationships, as aboriginal people have become less dependent on the land. Letendre (2002) explained that with large numbers of Europeans gaining control over the land base and its resources, Aboriginal people were forced to adjust from a lifestyle that was governed by laws of nature to a lifestyle governed by economy.

Other concepts particular to the Aboriginal community include an orientation towards group unity and cohesiveness, non-interference, non-competitiveness and emotional restraint and sharing, all considered to be essential for one's survival in hostile environments. The concept of non-interference is regarded as a promotion of positive interpersonal relations as it discourages coercion of any kind, be it physical, verbal or psychological. An attempt to exert pressure by advising, instructing, coercing or persuading is seen as undesirable behaviour, as every individual's independence is highly respected. In Western society some might equate this with the principle of autonomy (Brant 1990). The practice of non-competitiveness is meant to suppress conflict by discouraging problems between members of a group. Additionally, it is meant to prevent the less-able or less-successful members from feeling embarrassed or inferior. Brant (1990) noted that this behaviour is often interpreted by non-Aboriginal employers as a lack of initiative and ambition among their aboriginal employees.

Because group survival was so critical, as a way of life individuals were expected to take only as much as they needed from nature so that there was enough for others. Sharing not only discourages the hoarding of resources for it also reduces the likelihood of greed, envy, arrogance and pride. In addition, sharing promotes economic and social heterogeneity and exceptions are only made for the elderly population. These concepts are reflected in current commentaries on being 'in country' as being associated with traditional values of communal living (Degnen 2001). In recent years, the revival of annual gatherings and community healing events 'in country' have been described by anthropological observers and their informants as reviving and emphasizing more fluid and reciprocal relationships, and about being together in time and space (Scott 2001; Shuurman 2001).

A number of writers have argued that the Inuit perception of time also differs significantly from non-Inuit societies (Briggs 1992; Halpern and Christie 1990; Mauss 1979, Stern 2003). Nisbet (1997) for example, argued that Inuit traditionally measured distance and time not in miles, but in 'sleeps'. In other words, one would measure time and distance by calculating how many stops for rest were necessary to get from one place to another. Everyone understood that this measure could be affected by weather, topography, luck, skill and the number or age of the travellers. Whidden (1981) argued that the overall difference in time concepts evidenced in music and Inuit life may be described as cyclical versus linear. The traditional songs are cyclical in a time sense. However, the complexity of everyday life for today's Inuit has increased with an unprecedented rapidity. Although one of the elements that has enabled Inuit to survive the rigors of life in the Arctic has been their incredible ability to adapt, the speed and degree of change in society has taken its toll on the both people and culture.[3]

The geographers Goehring and Stager (1991) argued that the Inuit of northern Canada have undergone great changes in concepts of time and space as a result of the enormous changes brought about by colonial contact and modernity, in which traditional conceptions of time and space have been readjusted to meet the demands of externally imposed industrial time. They argued that settlement in villages and participation in the cash economy, plus the use of a range of modern goods and media, have brought about dramatic time-space compression. Nonetheless, echoing the classic work of Mauss on seasonal variations in temporal regulation (Mauss 1979), they noted how people continue to respond to the extreme seasonal changes of the region, particularly the long daylight hours heralded by spring, according to more traditional norms. Having survived *ubluiq* ('time of blackness') people begin to enjoy *arviktuk* ('time of light'):

> Time and space at this place cease to retain the linear meaning of lower latitudes. Inuit hunters and their families leave the settlement whenever they are ready, providing conditions are right for sea-ice travel. People come and go at all hours of the day. Men routinely hunt seals for 20, 30 or more hours at a time; women prepare food for their families when they get hungry, and everyone sleeps when they are tired ... Time becomes endless, and were it not for the artificial constructs of clock and calendar, essentially meaningless (Goehring and Stager 1991: 667).

From their account it is apparent that the passage of time is not meaningless, since people seize the moments of this season of light, but the modern, conventional marking of time is clearly less salient. Similarly, people wear wristwatches and use speedometers but the values they assign to concepts of time differ, creating what Goehring and Stager call 'an everyday crisis of meaning' (1991: 668). They suggested that the importance of *timing* in such an extreme environment was contrasted with a sense of *time* as eternal. In an oral culture, historical time was marked by generations (and so with reference to the continuity of social relationships) and space was marked in terms of considerable practical and symbolic relationships with the landscape. They noted that dramatic changes in people's life-world have taken place in the space of one or two generations, echoed in a shift from what they referred to, following Jackson (cited in Goehring and Stager 1991:673) as primarily polychromatic to monochromatic notions of time. Like anthropological writers, they argued that such cultural changes resulting from colonialism and the postcolonial experience of modernity, are at the heart of the identity crisis felt by many Aboriginal communities in the late twentieth century.

The anthropologist Stern (2003) also argued that the experience of modernity has meant changes in time discipline, although she cautioned against assuming that all concepts were opposed, or that the people have simply been passive recipients of change. She described the considerable turnabout between winter and summer time, or 'backwards time', that has been commented on by a succession of anthropologists and is now marginalized by the new norms of time discipline. She argued that although traditional values of being active, and using time well, are not at odds with modern time disciplines, 'modern' time concepts tend to discourage participation in traditional subsistence activities, which do not fit with the daily time regulation of wage labour. Although traditional notions of time might be described as more fluid, she noted that these also had ecological and social dimensions that gave importance to the marking of time. Traditional calendars were based on lunar time and seasonal variations of solar time, but the marking of weeks did not occur and daily patterns were subject to great seasonal and socially governed variation. A number of developments, including the introduction of Christianity, the move to settlements and wage labour, and the introduction of broadcast media with fixed time schedules, have influenced time notions and discipline, such as those of clock time, fixed-hour daily schedules, the division of working days and weekends and so on. Stern also noted the impact of changing norms and disciplines of time on reproduction – via such things as the scheduled nature of contraceptives such as the pill, and changing patterns of work – with a consequent effect on birth spacing and the seasonality of childbirth.

James (2001) warned against problems of 'essentializing' or reifying culture, treating it as though it is something fixed and unchanging. She suggested, therefore, that caution needs to be used in focusing on well-known ideologies and symbols such as cyclical versus linear concepts of time, egalitarian versus hierarchical societies and so on. She argued that Aboriginal cultures should not simply be defined in oppositional or dichotomous terms – that is, in relation to differences between them and the White majority – since this may simply add to a sense of dependency, and she noted that in areas such as attitudes to reproduction there are changes and continuities in attitudes. Nonetheless, the literature discussed above illustrates well the degree of impact that cultural change has had on remote northern communities, as well as the continuities to be found, and the resilience with which people are responding today to the many problems they face. The case in point here – continuity and change in maternity services – illuminates these issues well.

## Maternity Care and Midwifery in Canada

Maternity care is a publicly funded health service in Canada (Health Canada 2005). It begins with preconception care, continues with prenatal care and education and concludes with services to mother and infant until six weeks after birth. The range of services is delivered by a variety of healthcare professionals (from public health, acute care and community health units); lab and diagnostic imaging services; transport teams; breastfeeding, neonatal and child health programs, among others.[4]

Most babies are born in hospital with a physician as the attending clinical professional, but although the number of hospital births attended by midwives is increasing, midwifery care in Canada remains far from becoming a mainstream service (CIHI 2004). In 2002, in Canada only 2 per cent of total births were attended by midwives; in the same year in England midwives attended 72 per cent, and in New Zealand 70 percent of total births were attended by midwives.[5]

In recent years, Canada's maternity care environment has been in crisis. The shortage of maternity-care providers, including family physicians, obstetricians, nurses and midwives, is a well-documented fact.[6] In rural and remote regions of the country in particular, maternity care services are in decline. Between 2002 and 2004, thirteen of the sixty-two rural maternity services in British Columbia stopped providing birthing services to their communities (Kornelson and Grzybowski 2005), due to the nationwide shortage of obstetrical care providers and the centralization of services due to regionalization and system restructuring.[7] Klein (2002) suggested that these changes possibly led to serious health and social costs for birthing women and their families, and taxed health system resources. In communities where maternity services have been lost, women are sent to larger centres to deliver their babies in regional hospitals. Transfer out of the community for birth is costly and contributes to increased financial and social burden to the family. The loss of maternity capability in the community also increases complication rates and obstetrical interventions, contributing to escalating health care costs.

Northern and Aboriginal women have to travel weeks, sometimes months, in advance and over long distances outside their communities to access intrapartum services (Becker 2006). Most often they leave their spouses at home to look after their other children. Parturient women are often forced to experience birth without the presence and support of family members and community. The removal of birthing support from the communities has turned childbirth into a stressful event that disrupts rather than strengthens families and communities. Transfer out of the community for birth is costly and means an increased financial burden on the family. Rurality and remoteness are powerful determinants of women's health, as both a geographic and socio-cultural influence. Women living in rural and remote communities are largely invisible to policy makers and women's views on healthcare issues are mostly unknown (Suthern, McPhedran and Haworth-Brockman 2004).

Even though Aboriginal women have been practising as midwives in Canada since pre-colonial times, and immigrant women have practised likewise since the beginning of European colonization, only in recent times has midwifery legislation been introduced and the profession become part of the established healthcare system.[8] Midwives played a vital role in early Canada, but their work in families and communities was eventually replaced with care given by doctors and nurses in hospitals (Benoit and Carroll 2001). The history of Canadian midwifery includes: Aboriginal midwives who used a formal apprenticeship model; traditional lay midwives, who date back to the early settlement period;

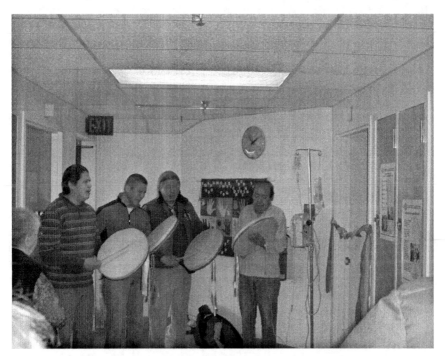

Figure 7.1: Drummers at the birth centre opening.

nurse midwives who received training outside the country; and certified midwives, whose numbers declined in Canada during the nineteenth century due to the Canadian Medical Act and the establishment of the hospital system (Benoit and Carroll 2001). By the late nineteenth century, birth was gradually redefined in more medicalized terms and the role of midwives was undermined, ultimately leading to their declining presence in the birthing process and, in many places, to their elimination altogether. For many years Canada was one of nine countries which did not recognize midwifery, and even today there are several provinces and territories where the profession is not regulated.[9]

Today, midwifery is a rapidly growing profession in Canada. Legislation and models of midwifery care are diverse across the country. The status of midwifery in many provinces and territories is dynamic and the process of legislating midwifery is moving quickly across Canada.[10] Meanwhile, in those parts of the country where midwifery is not legislated, consumer lobby groups are working towards the establishment of midwifery as a funded, self-regulating profession.

## Aboriginal and Northern Midwifery: Past and Present

Before the arrival of a non-Aboriginal, southern healthcare system in the north and in Aboriginal communities, women gave birth as they lived, on the land and in small communities with the help of family members and traditional midwives. Medical anthropologist John O'Neil described the traditional birthing practices of the Inuit in the Keewatin Region: 'Prior to 1965, nearly all Inuit babies in the Keewatin Region of the Northwest Territories were born in tents or snow houses, assisted only by members of the family and traditional midwives' (O'Neil, Kaufert and Postl 1990: 6). Aboriginal people taught women at a young age to acquire the skills and customs related to conception, pregnancy and birth. In the exhibition 'Midwifery in Canada', held at the Museum of Civilization, Benoit and Carroll (2003) explained that midwifery knowledge was passed down from generation to generation and required no sanction from external medical educational processes or legislation.

In the wake of rapid culture change and the growth of medical services in the mid twentieth century, traditional birthing practices were gradually displaced and traditional midwifery disappeared in many places. Benoit and Carroll (2001) identified a number of significant historical events that led to the demise of Aboriginal midwifery, including colonialism and Western medicine. The introduction of government legislation and policies changed the traditional ways of Aboriginal people and undermined their long-standing healing traditions.

During the 1960s and early 1970s, maternity care outside the larger centres was provided primarily in nursing stations with the aid of nurse-midwives, many of them trained in the United Kingdom (Paulette 1990). Although community birthing was still commonplace, the end of the nurse-midwives period seemed inevitable. Since the mid 1970s, policies in maternity care have changed, either due to difficulties in hiring foreign-trained nurse-midwives to work in remote northern communities, or to changing attitudes about acceptable risk levels in maternity care. Current policy dictates that all expectant mothers are to travel from smaller communities to give birth in larger regional centres (Paulette 1990). At present, most women are forced to leave home for several weeks, sometimes months, to deliver their babies in regional hospitals under physician care, without the presence and support of partners, families and community members. This policy causes considerable disruption to the lives of families and communities and many women have described poor treatment and care in hospital settings which are, for them, remote (Browne et al. 2000).

Aboriginal and northern midwifery-led birth centres were established in direct response to intense lobbying from women, families and communities who were unwilling to accept the removal of local birthing support. The very first programme started in Puvirnituq in Nunavik in northern Québec in 1986. Later, the Innuulitsivik Health Centre opened two additional midwifery birth centres on the Hudson Bay Coast at Inukjuak in 1998 and Salluit in 2004. The Maternal and Child Centre of the Six Nations of the Grand River in southwest Ontario opened in 1996 and the birth program in Rankin Inlet in Nunavut began in 1993. And in 2005, a midwifery program was launched at the Fort Smith Health Centre in the Northwest Territories. Although the birth centres are to be found in different cultural settings and operate in their own unique ways, the sites share remarkable similarities. All of the described programs have reintroduced local birthing services with midwives so that women can birth in their home communities. Evidently the programs promote and provide culturally safe midwifery care. Cultural safety relates to the experience of the recipient of midwifery service and extends beyond cultural awareness and cultural sensitivity.[11] Cultural safety describes effective midwifery practice determined by the woman and her family, and enables safe, appropriate and satisfactory care that has been defined by those who receive the services.[12]

Midwifery training programmes have been a fundamental component of the centres in northern Québec and Ontario and are crucial to the development and maintenance of local capacity (Tedford Gold, O'Neil and Van Wagner 2005). Rankin Inlet has not offered midwifery training until recently, though a pilot project for an Inuit midwifery education program is currently underway in Nunavut; the curriculum is based on Inuit cultural knowledge of birthing.[13] The Northwest Territories remain without a midwifery education program in 2009, though the midwives at the Fort Smith Health Centre envision a local training program in the future.

Community birthing in remote communities occurs primarily in small centres without immediate surgical back-up available; emergency air-evacuation time to the next larger centre may take several hours. There have been extensive concerns about, and even opposition to, community birthing, with the argument that this practice may not be medically safe and that mothers and babies are at increased risks of adverse outcomes. However, the programs have successfully provided evidence that birthing in remote communities without immediate surgical back-up is safe. Furthermore, populations served by rural and remote centres which do not offer maternity care seem to have worse perinatal outcomes (Iglesias and Hutten-Czapski 1999).

In the following interviews with mostly Aboriginal (mainly Inuit) midwives, the midwives shared their thoughts and ideas about the management of time in Aboriginal and northern midwifery settings. It was vital to the midwives to provide women-focused and family-centered maternity care in their home communities. The midwives described the challenges around finding a balance in maternity care and childbirth between the traditional perception of time and the medical model's concept of time, which is also that of modern life.

## The Rankin Inlet Birthing Centre and Midwifery in Nunavut

The birthing centre in Rankin Inlet was established in 1993 in response to the desire of Inuit women to deliver their babies closer to home and within their own cultural context. Prior to the program there was clear evidence of substantial dissatisfaction and unhappiness with the existing obstetrical care among Inuit women in the Kivalliq region. Preference for community births and Inuit midwives as maternity care providers was high and evenly distributed throughout the population (O'Neil, Kaufert and Postl 1990).

For a number of years the birthing centre ran as a pilot project, and it was subsequently evaluated by Chamberlain et al. who aimed 'to identify whether community birthing would be safe, cost-effective, and psychologically and socially satisfying for Inuit women in Rankin Inlet' (1998: 118). Two midwives provided antenatal and postnatal care to all women and assisted the births of women who were designated as being at 'low risk' of complications. Another Inuit community similar in size but with no community birthing services was used for comparison. Health staff, community members and pregnant women and their partners in both communities were interviewed about their thoughts and feelings concerning the birthing services. The women who had had their infants in the community expressed satisfaction for a number of reasons. Women from Rankin Inlet felt that they participated in the decision of where birth should take place. For those women who delivered away from home labour appeared to be quite a traumatic event. This was intensified by feelings of isolation and worries for the families back home.

Today, the midwives working in the Rankin Inlet Birthing Centre in Nunavut provide maternity services for women and families across the

Kivalliq region. Over the years, the recruitment and retention of midwives has been taxing and this has repeatedly jeopardized the provision of maternity services and community birthing. Northern maternity services are particularly burdened by high travel costs, poor staff recruitment and retention, lack of local training programs and a lack of local midwifery services. When responding to demands for local, midwifery-based maternity care, decision makers refer to uncertainties around community participation, sustainability, legislation, standards and training.[14]

More recently, the government of Nunavut finally committed to midwifery legislation and midwifery education in the region.[15] Not long ago, a pilot project for an Inuit midwifery education program was established in Rankin Inlet, and its curriculum is based on Inuit cultural knowledge of birthing and maternity care. The first midwives graduated in 2007.[16]

### A Rankin Inlet Midwife

In an interview with Melanie Shields, one of the midwives working in Rankin Inlet, she described her perception of time as it relates to her practice experience in the Birthing Centre.[17]

> Midwifery practice in Rankin Inlet is altogether less rushed, also because life in general happens slower in a remote northern community. There is a lower volume of clients and [so] maternity care [for] women and family becomes more personal and individual, based on the needs of each woman and her family. Whatever time it takes to look after a woman, there is no time limit, such as 'you have to be done in thirty minutes'. Rankin Inlet is a small facility with no caesarean section back-up, several hours of flying time away from the nearest surgical centre, and [then] only if the weather conditions are in favour and the medevac plane is in the community. Otherwise, transport times may be much longer. It definitely influences your practice and your perspective of time. In general, the people and local community are quite laid-back about transport times if a woman or a baby has to leave the community for medical reasons. After all, people do not only birth in the community, they live here, and therefore people face this problem every time a medical condition presents itself and exceeds the capacities of the local healthcare facility. Because of the delay in accessing certain types of services in a reasonable time, every now and then a problem resolves by itself, and at another time it means trouble.[18]

Despite these remarks, the midwives' work schedule in the birthing centre is not flexible enough to accommodate a satisfying midwifery practice around the clock. Midwives are required to work regular working hours from 8.30 A.M. to 5.00 P. M. They are additionally on call after hours as per the agreement made between their union and the government. The lack of flexible time management has contributed to burnout and the exhaustion of midwives as many babies are born outside regular working hours. A better time management scheme with more flexible working hours would be desirable.

In commenting on the politics of local health care, both Lavoie (2001) and Shuurman (2001) noted the challenges and conflicts faced by local people in remote northern settings attempting to reintroduce locally controlled health services. However, time and work disciplines which have been imposed externally are not always contested because they have come to seem normal and natural to those concerned (Stern 2003).

## *Inuit Midwifery in Nunavik*

From the 1950s onward, women in Nunavik were transferred to southern Québec or Ontario to give birth (Houd 2003). Midwifery services started in 1986 at Innuulitsivik Health Centre in Puvirnituq, in the Nunavik region of northern Québec. The birthing movement was inspired by the community itself and developed through the hard work of Inuit women's organizations, associated with Inuit cultural revival, and is seen as part of the process of 'healing' from the effects of colonialism (Nastapoka and Van Wagner 2005). The Centre is governed by an Inuit community board and is committed to community development and the education of Inuit health workers (Lavoie 2001), who are mostly taught through 'on the job' learning. By 2006, nine midwives had graduated through the Inuulitsvik education process, and seven student midwives were currently learning. Traditional midwives and traditional knowledge about birth are part of the education program.

Today, the Innuulitsivik Health Centre operates three midwifery maternity centres on the Hudson Bay coast at Puvirnituq, Inukjuak and Salluit. Women from the remaining villages on the Hudson Bay coast are transferred to Puvirnituq at thirty-seven to thirty-eight weeks gestation, a situation which they feel is far easier and more acceptable than having to go to southern Québec, especially as the care they receive at the Innuulitsivik Centre is culturally safe and closer to home (Nastapoka and Van Wagner 2005). The Centre has achieved perinatal mortality and morbidity rates comparable to or lower than the rates for southern Québec.[19]

## *An Inuit Midwife at Innuulitsivik Maternity Centre*

In an interview with Mina Amamatuak, Inuit Midwife at the Innuulitsivik Maternity Centre in Puvirnituq, she described her perception of time as it relates to her midwifery practice and her understanding of Inuit culture.

Women giving birth at the Innuulitsivik Maternity Centre receive care from Inuit midwives in Inuktitut [the Inuit language]. Maternity care is culturally appropriate and the time is spent with women to address their needs in the best way. Sometimes, a visit may take longer than other times, depending on the situation and the circumstances. When a woman gives birth, the midwife does not watch the clock and say that this is taking too long ... the midwife does not rush the mother to give birth. Whatever time it takes for her to birth is acceptable. In the south, women are left alone in labour; this is incomprehensible in the Inuit culture. At the Innuulitsivik Maternity Centre, women are supported by their families and the community. Because care is provided by Inuit to Inuit, time is not wasted with things that have no meaning to Inuit people.

When a woman has to go south, often weeks in advance, time away from home, the community, children and family can be very lonely. When women stay closer to home, they do their work and daily routine until they give birth, time is not very long. Giving birth takes only a little while and it does not make sense to be sent out for weeks unless there is a medical condition.

Inuit midwives existed well before the midwives law came into force in the south. The community and Inuit women want the maternity ward to be run by midwives. Reconciling traditional Inuit practice with the best of modern medicine works for the midwives at the Innuulitsivik Maternity Centre![20]

## The Maternal and Child Centre of Six Nations of the Grand River

The Maternal and Child Centre of Six Nations of the Grand River in south-west Ontario provides comprehensive maternity services to women and their families, including preconception services, pre- and postnatal care, and birthing services to women with low-risk pregnancies.[21] Aboriginal midwives and support staff provide a balance of traditional and contemporary midwifery services and maternity care that incorporates traditional practices.[22] The Aboriginal community, family, and specifically the expectant woman, is offered a choice of services that complement and support personal beliefs and customs.

The Aboriginal midwifery practice of Six Nations operates within a community-driven structure that provides support and direction for the services that are provided to Aboriginal families in the communities.[23] Ontario midwifery legislation includes an exemption clause for 'an Aboriginal person who provides traditional midwifery';[24] practitioners may therefore use the title 'Aboriginal midwife' and they are not required to register with the College of Midwives of Ontario (see CMO 2001). The need for the service was identified by key community members, and supported by a 'Birth Search Conference and Needs Assessment', which substantiated 'the need for Aboriginal Midwives and for the community to regain control of the birth of our people' [25]. Once the centre was established, midwifery training of community women commenced and midwifery services were provided to community women and their families. The first class of Aboriginal midwives graduated in 2003.[25]

### An Aboriginal Midwife of the Six Nations of the Grand River

In an interview with Laurie Jacobs, an Aboriginal midwife of the Six Nations of the Grand River, she described her perception of time as it relates to her traditional and contemporary midwifery practice and her understanding of customs and beliefs.

Aboriginal women have many other issues besides pregnancy and birth, and if time were restricted it would be impossible to support women and their families in a good way. Because aboriginal people have 'higher rates of everything', a different approach is necessary to provide maternity care. This includes the provision of holistic care and gaining the woman's and her family's trust. It is important to create a relaxed atmosphere, where there are no time limitations. The Maternal and Child Centre also provides transportation from the woman's home to the centre and sometimes back again. It is when the woman talks about things important to her and when good conversations happen.

Midwives are part of the Aboriginal community and frequently engage with women in conversations and other activities in informal settings; again, time does not matter. Aboriginal midwifery care is not an assembly line and whatever time is required to be with the woman will be the time spent with her. Aboriginal people live more in 'Indian time', and women may not come to their appointments or may be late for their appointments, not because they do not care about it, but because the appointment may not have significance at this particular moment or it may have been simply forgotten because of other tasks. The Maternal and Child Centre educates midwives, and working with students takes a lot of time. There are also many administrative responsibilities to be completed, which can be very time consuming. The midwives and staff in the Maternal and Child Centre work hard; it is necessary to have enough time away from work to maintain a balance between work and family life. Midwives need to take care of themselves as well![26]

## The Midwifery Program at the Fort Smith Health Centre

The Midwifery Program at the Fort Smith Health Centre was established in 2005. Currently, two midwives are employed by the Fort Smith Health and Social Services Authority. The midwives provide comprehensive maternity services to women and their families, including preconception services, pre- and post natal care, and birthing services to women with low risk-pregnancies. The midwives are also involved in the provision of well-woman care and health promotion activities. The midwives are members of the Maternity Care Committee of the Health Authority. The interdisciplinary committee reviews client care and makes recommendations in regard to birthplace choices and maternity care.[27]

For a number of reasons, the Fort Smith community played a significant role in bringing about midwifery legislation to the Northwest Territories. Initially there was one Aboriginal midwife of Mohawk descent in the community who had lived in the Fort Smith area since 1983,[28] and whose interest and training in the field of midwifery dated back to 1981. She established her practice in Fort Smith in 1992 and has provided midwifery services to local families since then. Through her efforts, community birthing services have returned to Fort Smith with the result that grassroots support from a growing number of women and families in Fort Smith has grown. The local health authority had expressed interest in the provision of midwifery services but had identified the absence of legislation as a barrier (GNWT 2003). However, the Minister of Health took an interest in the midwifery issue, lending political support to the movement to recognize the midwifery profession.[29] Most recently, midwifery legislation has been passed in the Northwest Territories, which provides the legal framework to introduce regulated midwifery services to other communities of the Northwest Territories.

### A Midwife at the Fort Smith Health Centre

In an interview with Lesley Paulette, an Aboriginal midwife of Mohawk descent, she described her perception of time as it relates to her midwifery practice and her cultural beliefs.

Today, the concept of time in childbirth is imposed from the outside, and time constraints are more rigid than twenty to thirty years ago: for example, long labours or going past the due date. When a woman goes past her due date, the midwife has to practice 'evidence-based' midwifery and inform her client that an induction is recommended at forty-one weeks gestation because of a slightly increased risk. The midwife is caught up in the dilemma of practising this kind of midwifery and how Aboriginal people would have handled this traditionally. Practising as an Aboriginal midwife there is definitely a conflict in value systems.

The time has changed the medical climate and there is also the fear of negative pressure and scrutiny form the medical world if a situation is handled in a traditional way. Traditionally, the Aboriginal midwife promotes physical activity, a balanced diet and staying healthy during pregnancy. However, today the midwife has less influence as women may not listen in the same way to the midwife as they used to.

Annie Catholique was a highly respected Dene midwife from Lutselk'e who learned her skills from her mother. She delivered babies under the most challenging of conditions, handling everything from normal, straightforward births to breech births, very premature births and the manual removals of placentas. Annie told Lesley that she knew intuitively when it was time for a baby to arrive, and when the time was right. This is different from the medical model.

There is also a noticeable shift in the community, and giving birth within it has disappeared in most places. People are not used to having pregnant women at term in the community and are not comfortable with women who have gone past their due dates. There is a sense of impatience in the community and pressure is put on women to give birth without delay. It continues to be a challenge to find a balance between traditional time perception in childbirth and the time concept of the medical model.[30]

# Discussion

In the interviews reported above, the midwives described the particular experiences of time management in a variety of Aboriginal and northern midwifery settings. One midwife spoke about a generally slower pace of life in small communities, where midwives have more time for each woman, and where there are long journey times out of the community and exhausting working hours. Another midwife described the significance for Inuit women of being cared for by Inuit midwives, where time is spent with women addressing their needs in a culturally safe way. In Inuit culture women are not rushed to give birth: whatever time it takes for her to give birth is acceptable. A different midwife said that having no time restrictions when caring for women and their families is most important when trying to gain their trust and make a difference in people's lives. It is also important to spend time with women informally in the community, and time in its modern sense should not matter. As midwives work hard it is vital for them to spend time away from work to enjoy time with their families, friends and community. One midwife spoke about the role of traditional Aboriginal midwives and how today they have less influence as women may not listen in the same way as they used to.

The same midwife pointed out that the concept of time in childbirth is imposed from the outside, and that the time constraints are more rigid today. This echoes the arguments of a number of anthropologists who have discussed the ways in which modern time disciplines have been not just imposed but contested and used creatively within northern and Aboriginal communities (Stern 2003). The historian Kelm (1998) similarly argued that women initially welcomed free access to hospital services for childbirth, seeing it as a 'rights' issue, and because many health problems the communities experienced were associated with modernity and non-native contact, women felt they should be dealt with by modern means.

The midwives did not view the concept of time as an abstract theory, and instead they spend time with women during pregnancy, childbirth and the postpartum period. The midwives felt that 'Aboriginal time' facilitates women-centred care that is responsive to the needs of the woman and her family, including other aspects of life besides pregnancy and childbirth. This includes the ability to give birth in the home community, to receive care from midwives of a woman's own culture and to receive care in a culturally safe manner, to build a partnership with midwives, and to spend as much time as necessary with them.

The interviewed midwives acknowledged that they have to juggle Aboriginal time perception in childbirth and the time concept of the medical model. In addition, Aboriginal midwives face other challenges that are imposed by the outside, such as increased work loads, time constraints and changing values within their own cultures. Not only does the concept of time affect maternity care for women and babies in Aboriginal and northern midwifery settings, as other factors – such as remoteness, socio-cultural impact, political control, the accessibility of services and conflicting values – significantly influence childbirth as it relates to time.

## Notes

1. Ontario Women's Health Council. 2002. 'Ontario Women's Health Status Report'. Retrieved 30 December 2005 from: http://www.turtleisland.org/healing/women1.pdf.
2. Grace, S.L. 2002. 'Aboriginal Women', in Ontario Women's Health Council. 2002. 'Ontario Women's Health Status Report'. Retrieved 30 December 2005 from http://www.turtleisland.org/healing/women1.pdf
3. Qitsualik, R.A. 1999. 'Living with Change: Nunavut 1999'. Retrieved 31 December 2005 from http://www.nunavut.com/nunavut99/english/change.html.
4. Ontario Maternity Expert Panel. 2006. 'Executive Report of the Ontario Maternity Expert Panel, Emerging Crisis, Emerging Solutions'. Retrieved 10 March 2007 from: http://meds.queensu.ca/prn/pdfs/Executive%20Report%20FINAL.pdf.
5. British Columbia Centre of Excellence for Women's Health. 2003. 'Solving the Maternity Care Crisis'. Retrieved 10 March 2007 from: http://www.bccewh.bc.ca/policy_briefs/Midwifery_Brief/midwifbrief%20v4.pdf.
6. See the 2005 report of the Multidisciplinary Collaborative Primary Maternity Care Project (MCP). Retreived 1 April 2006 from: http://www.mcp2.ca/english/welcome.asp.
7. Rural Maternity Care Research. 2005. 'Summary of the State of Rural Maternity Care in British Columbia'. Retrieved 10 March 2007 from: http://www.ruralmatresearch.org.
8. Herbert, P. 2006. 'A Summary of the History of Midwifery in Canada'. Association of Midwives of Newfoundland and Labrador. Retrieved 3 December 2006 from: http://www.ucs.mun.ca/~pherbert/Historyofmidincanada.html.
9. Source: see note 7.
10. Canadian Association of Midwives. 2006. 'Across Canada'. Retrieved 3 December 2006 from http://www.canadianmidwives.org/across_canada.htm.
11. Nursing Council of New Zealand. 2002. 'Guidelines for Cultural Safety, the Treaty of Waitangi and Maori Health in Nursing and Midwifery Education and Practice'. Retrieved 26 October 2004 from: http://www.nursingcouncil.org.nz/culturalsafety.pdf.
12. Nursing Council of New Zealand. 2005. 'Guidelines for Cultural Safety, the Treaty of Waitangi and Maori Health in Nursing Education and Practice'. Retrieved 11 February 2006 from http://www.nursingcouncil.org.nz/Cultural%20Safety.pdf
13. Canadian Association of Midwives. 2005. 'Across Canada – Nunavut'. Retrieved 2 January 2006 from: http:canadianmidwives.org/Nunavut.htm.
14. Tedford Gold, S. and J. O'Neil. 2002. 'Canadian Health Research Foundation. Examining Midwifery-based Options to Improve Continuity of Maternity Care Services in Remote Nunavut Communities'. Retrieved 2 January 2006 from: http://www.chsrf.ca/funding_opportunities/ogc/2002/tedford_e.php.
15. Legislative Assembly of Nunavut. 2002. 'Hansard, Fifth Session. Moving Towards Midwifery Training in Nunavut'. Retrieved 2 January 2006 from:

http://www.assembly.nu.ca/old/english/hansard/final5/020228.html#
b16015.

16. On this project, see: Canadian Association of Midwives. 2005. 'Across Canada
    – Nunavut'. Retrieved 2 January 2006 from: http://www.canadianmidwives.
    org/nunavut.htm.

17. The interview notes reproduced in this chapter come from interviews which
    were not tape-recorded but at which notes were taken. Thus, although
    the notes do not represent verbatim quotations, they do represent the
    interviewees' own words.

18. Personal interview with the author, 18 December 2005.

19. National Aboriginal Health Organization. 2004. 'Midwifery and Aboriginal
    Midwifery in Canada'. Retrieved 4 January 2006 from: http://16016.vws.
    magma.ca/english/pdf/aboriginal_midwifery.pdf.

20. Excerpt from personal interview with the author, 16 December 16 2005.

21. Aboriginal Healing and Wellness Strategy. 2003. 'Programs and Services:
    Maternal and Child Centre on Six Nations of the Grand River'. Retrieved 8
    January 2006 from: http://www.ahwsontario.ca/programs/maternal.html.

22. National Aboriginal Health Organization. 2003. 'Six Nations Maternal and
    Child Centre'. Retrieved 7 January 2006 from: http://www.naho.ca/english/
    pdf/ABirthingPlace-SixNations.pdf.

23. Source: see note 20.

24. Act Representing the Regulation of the Profession of Midwifery (Ontario
    1991), p. 3

25. National Aboriginal Health Organization. 2003. 'Six Nations Maternal and
    Child Centre'. Retrieved 7 January 2006 from: http://www.naho.ca/english/
    pdf/ABirthingPlace-SixNations.pdf.

26. Personal interview with the author, 21 December 2005.

27. Fort Smith Health and Social Services Authority. 2005. 'Terms of Reference'.
    Maternity Care Committee, Fort Smith, Northwest Territories.

28. ChildEarth Resources. 2002. 'Midwives Serving Women and Families in Fort
    Smith'. Information Brochure, Fort Smith, Northwest Teritories.

29. Hansard. 2001. 4th Session, 14th Assembly. Legislative Assembly, Government
    of the Northwest Territories. Yellowknife, Northwest Territories.

30. Personal interview with the author, 24 December 2005.

# References

Becker, G. 2006. 'Women's Experiences of Culturally Safe Birthing with a Midwife
    in a Remote Northern Community'. M.A. dissertation. London: Thames
    Valley University.

Benoit, C. and D. Carroll. 2001. 'Aboriginal Midwifery in Canada: Blending
    Traditional and Modern Forms', *Network Magazine* 4(3): 6–7.

———— 2003. *Midwifery in Canada*. Hull, Québec: Museum of Civilization.

Brant, C. 1990. 'Native Ethics and Rules of Behaviour', *Canadian Journal of
    Psychiatry* 35: 535–39.

Briggs, J. 1992. 'Lines, Cycles and Transformations: Temporal Perspectives on Inuit
    Action', in S. Wallman (ed.) *Contemporary Futures,* London: Routledge,
    pp. 83–108.

Browne A.J., J. Fiske and G. Thomas. 2000. *First nations women's encounters with mainstream health Care services & systems*. Vancouver, B.C. : British Columbia Centre of Excellence for Women's Health, 2000.

Chamberlain, M. et al. 1998. 'Evaluation of a Midwifery Birthing Centre in the Canadian North', *International Journal of Circumpolar Health* 57(Suppl. 1): 116–20.

CIHI. 2004. 'Giving Birth in Canada', in *Providers of Maternity and Infant Care*. Ottawa: Canadian Institute for Health Information.

CMO. 2001. *Registrant's Binder*. Toronto: College of Midwives of Ontario.

Degnen, C. 2001. 'Country Space as a Healing Place: Community Healing at Sheshatshiu', in C.H. Scott (ed.) *Aboriginal Autonomy and Development in Northern Quebec and Labrador*. Vancouver: University of British Columbia Press, pp. 357–78.

Goehring, B. and J.K. Stager. 1991. 'The Intrusion of Industrial Time and Space into the Inuit Lifeworld: Changing Perceptions and Behaviour', *Environment and Behavior* 23(6): 666–79.

GNWT. 2003. *Progress Report*. Yellowknife, NWT: Government of the Northwest Territories.

Halpern, J.M. and L. Christie. 1990. 'Temporal Constructs and "Administrative Determinism": A Case Study From the Canadian Arctic', *Anthropologica*, 32: 147–65.

Health Canada. 2005. *Canada's Health Care System*. Ottawa: Health Canada.

Houd, S. 2003. 'The Outcome of Perinatal Care in Inukjuak, Nunavik, Canada 1998–2002', *Birth International* 63 (Suppl. 2): 239–41.

Iglesias, S. and P. Hutten-Czapski. 1999. 'Joint Position Paper on Training for Rural Family Practitioners in Advanced Maternity Skills and Cesarean Section', *Canadian Journal of Rural Medicine* 4(4): 209–16.

INA. 2000. *Comparison of Social Conditions, 1991 and 1996*. Ottawa: Indian and Northern Affairs, Minister of Public Works and Government Services.

James, C. 2001. 'Cultural Change in Mistissini', in C.H. Scott (ed.) *Aboriginal Autonomy and Development in Northern Quebec and Labrador*. Vancouver: University of British Columbia Press, pp. 316–31.

Kelm, M.E. 1998. *Colonising Bodies: Aboriginal Health in British Colombia, 1900–1980*. Vancouver: University of British Colombia Press.

Klein, M. 2002. 'Mothers, Babies and Communities', *Canadian Family Physician* 7(48): 1177–79.

Kornelson, J. and S. Grzybowski, S. 2005. *Rural Women's Experiences of Maternity Care: Implications for Policy and Practice*. Ottawa: Library and Archives Canada Cataloguing in Publication.

Lavoie, J. G. 2001. 'The Decolonization of the Self and the Recolonization of Knowledge: The Politics of Nunavik Healthcare', in C.H. Scott (ed.) *Aboriginal Autonomy and Development in Northern Quebec and Labrador*. Vancouver: Uuniversity of British Columbia Press, pp. 332–56.

Letendre, A.D. 2002. 'Aboriginal Traditional Medicine: Where Does It Fit?' *Crossing Boundaries – An Interdisciplinary Journal* 1(2): 78–87.

Mauss, M. 1979. *Seasonal Variations of the Eskimo: A Study in Social Morphology*, J.J. Fox (trans.), London: Routledge & Kegan Paul.

Nastapoka, J. and V. Van Wagner. 2005. 'Keynote Speeches at the Brisbane Congress from Canada, Malawi, the U.K. and Australia: Four Renowned Midwives Gave Inspirational Addresses on the Congress Themes of

History, Professionalisation, Current Ways of Knowing and Future Pathways.' *International Midwifery*.

Nisbet, B. 1997. 'Concepts of Time and Space in Inuit Art', *Journal of the International Institute* 4(2): http://hdl.handle.net/2027/spo.4750978.0004.217.

O'Neil, J.D., P. Kaufert and B. Postl. 1990. *A Study of the Impact of Obstetric Policy on Inuit Women and their Families in the Keewatin Region, NWT.* Winnipeg: University of Manitoba.

Paulette, L. 1990. 'The Family Centred Maternity Care Project' in M. Crnkovich (ed.) *Gossip: A Spoken History of Women in the North*. Ottawa: M.O.M, pp. 71–87.

Scott, C.H. (ed.) 2001. *Aboriginal Autonomy and Development in Northern Quebec and Labrador.* Vancouver: UBC Press.

Shuurman, H. 2001. 'The Concept of Community and the Challenge of Self-Government', in C.H. Scott (ed.) *Aboriginal Autonomy and Development in Northern Quebec and Labrador.* Vancouver: University of British Columbia Press, pp. 379–95

Siggner, A.J. 2003. 'Urban Aboriginal Populations: An Update Using the 2001 Census Results', in D. Newhouse and E. Peters (eds), *Not Strangers in These Parts: Urban Aboriginal Peoples*, Ottawa: Policy Research Initiative, pp. 15–21.

Statistics Canada. 1998. *1996 Census: Aboriginal Data.* Ottawa: Minister of Industry.

Statistics Canada. 2003. *Aboriginal People of Canada.* Ottawa: Minister of Industry.

Stern, P. 2003. 'Upside-down and Backwards: Time Discipline in a Canadian Inuit Town', *Anthropologica* 45:147–61.

Suthern, R., M. McPhedran and M. Haworth-Brockman. 2004. 'Summary Report: Rural, Remote and Northern Women's Health', Vancouver: Centres of Excellence for Women's Health, Policy and Research Directions. Available from: http://www.cewhcesf.ca/PDF/cross_cex/RRN_Summary_ CompleteE.pdf

Tedford Gold, S., J. O'Neil and V. Van Wagner. 2005. *Examining Midwifery-based Options to Improve Continuity of Maternity Care Services in Remote Nunavut Communities.* Ottawa: Canadian Health Research Foundation.

Whidden, L. 1981. 'Charlie Panigoniak: Eskimo Music in Transition', *Canadian Journal of Traditional Music*. Available from: http://cjtm.icaap.org/content /9/v9art4.html

# Part III

## TIME AND CHILDBIRTH EXPERIENCES

*CHAPTER 8*

# NARRATIVE TIME: STORIES, CHILDBIRTH AND MIDWIFERY

*Ólöf Ásta Ólafsdóttir and Mavis Kirkham*

Midwives and mothers have always told stories of births and their surrounding events and relationships. This chapter will explore the role of stories and storytelling in a piece of narrative research concerning childbirth and midwifery knowledge development. We will discuss themes from birth stories across time and space, mainly focusing on urban and rural Iceland and the U.K. This research is based upon work using ethnographic and narrative methodologies in relation to different aspects of midwives' relationships with women (Kirkham 1997a, 2004). It is designed from a broad perspective, exploring the storytelling of Icelandic midwives' in the period from the mid twentieth century to the present time, and it also focuses on the midwives' relationships with the women and those women's ways of knowing (Ólafsdóttir 2006).

Birth stories are all around us and they influence how we think about childbirth and about ourselves. We could say that we live in multiple stories that represent different experiences and views. They can be the official story contained in medical records or personal stories of women and families in society. They can also be stories of the working lives of midwives and doctors. These multiple stories form a web, which echoes the web of relationships and values within each story.

In a traditional society, Brigitte Jordan states that 'to acquire a store of appropriate stories and, even more importantly, to know what are the appropriate occasions for telling them, is part of what it means to become a midwife' (1993: 195–96). Very different stories are similarly used by modern professions to reinforce and teach their values and identity (see, e.g., Simpson 2004). Using appropriate research methods, we seek to examine the nature of birth stories, the knowledge they contain, and what they can demonstrate on particular occasions in terms of individual and professional identities (Hulst and Teijlingen 2001).

Stories, of course, are told to a particular end. This may be to enhance understanding but usually the aim is more specific and rhetorical. In this chapter, we seek to show what can be seen within stories and their potential uses, particularly with regards to time and childbirth.

## Storytelling and Narrative Research

Traditionally, midwifery was based on experiential rather than academic learning. Much of midwifery knowledge is therefore practical and taken-for-granted or tacit knowledge. The same can be said of women's knowledge about childbirth in general. Some of this knowledge comes from observation but much is distilled from learning from the stories of others.

It is difficult and challenging to try to access practical or traditional knowledge and to find an appropriate scientific way to identify and uncover it. According to Frank (2000), a narrative is a structure underpinning a story, and narrative analysis locates structures that storytellers rely on but of which they are not fully aware. As Wengraf (2001) points out, the advantage of the narrative approach, therefore, is that it conveys the tacit and unconscious assumptions and norms of the individual or of a cultural group. Take the following example, in which a midwife illustrates her position in the obstetrical culture of hospitals: 'You drop into the waiting-on mode [of doctors] before you've even realised you've done it. It's automatic to think they're top and you're bottom' (quoted in Kirkham 2004: 126). Values are similarly displayed in parents' stories.

Telling stories has been a way of learning in midwifery practice. Older midwives have been role models for younger ones and archetypal stories have been passed on for the messages they contain, making stories a mode by which midwifery knowledge is transmitted (Jordan 1993; Kirkham 1997a; Hulst and Teijlingen 2001). Recently, one of us (M.K.) observed an independent midwife preparing a couple, whose previous birth had been very difficult, complex and prolonged, for their planned home birth. The midwife told a story of a similar couple whose baby had arrived rapidly, before the midwife. Later, the midwife explained that she told the story for two reasons. Firstly, she wanted to demonstrate that women, in their chosen setting, can labour rapidly and effectively despite very different previous experiences. Secondly, she sought to prepare the father in case he had to deliver the baby. A fortnight later the baby was born at home, with the midwife in attendance. In later discussion both parents agreed that hearing the story was the point at which 'what we had planned for but hardly dared hope for started to feel real'.

Midwives' stories can also serve as rhetoric. Rhetorical analysis has been defined as 'the study of how people persuade (McCloskey 1985: 29). Such analysis, particularly when it is applied to stories told in conflict situations, pinpoints what the community sees as truth, 'what constitutes its authoritative knowledge about important subjects and what evidence the community uses to sustain or alter its knowledge' (Lay 2000: 12). Much can also be learnt from the context and intentions behind storytelling. Midwives may seek privately to enhance a woman's postnatal mental health by retelling her own story, emphasizing her bravery and endurance or, in a legal context, they may tell stories defending their profession by demonstrating the complexity and sensitivity of their practice (Lay 2000).

## Narratives: Definitions and Typology

A story or narrative contains a temporal sequence, a patterning of events. It has a social dimension, as someone is telling something to someone, and it has meaning, a plot giving the story a point and unity (Kvale 1996), plot being defined as 'a type of conceptual scheme by which a contextual meaning of individual events can be displayed' (Polkinghorne 1995: 7). Narratives also have an evaluative function.

Narratives are configurations that generate a story (emplotment) and the plot's integrating operation is the story. When the events or happenings of the story are configured, they take on narrative meaning as it is understood from the perspective of the storyteller and their aim in telling the story (Polkinghorne 1995). The storyteller tells the listener about events that are important to them and their evaluation makes a point. The resulting narrative provides new, convincing insights and opens new ways for understanding the diversity of childbirth experience as well as health professionals' practice and different models of care.

One of the main social functions of narratives is maintaining social ties (Kvale 1996). Sharing an experience, told in a story, can be strengthening for the teller and for listeners and entering the relations of storytelling is empowering for persons, relationships, and communities (Frank 2000). Frank (2000) maintains that stories are more than 'data for analysis' as storytelling calls for other stories in which experience is shared, commonalities discovered and relationships built. This can be seen between individuals and in formal educational settings such as midwifery education and antenatal classes.

Equally, the pared-down narrative of clinical records, in which only those facts that are accepted as medically relevant are included, is recreated with every birth. This reinforces the 'objectivity' of such a narrative, the irrelevance of other knowledge and the fault of the narrator (the clinician) if relevant data are missing.

As there are different types of stories and narratives, creating and retelling stories are different types of actions (Greenhalgh and Hurwitz, 1998). All discourse is underpinned with values and rhetorical intent. Sometimes conflicting values are transmitted. On the one hand, we have repeated narratives by midwives about birth being a normal family event or that 'breast is best' for the baby, while on the other we have medical narratives about the risks of birth, the dangers of homebirths or the need for a specific level of infant weight gain which may entail supplementary bottle feeds for breastfed babies. Modern society accepts a very medicalized view of birth and child development. It is therefore not surprising that such conflicting discourses are seen in the media, in official documents and retold by midwives.

With increasing technical interventions in Western childbirth care, alongside the continuing discourse of natural childbirth, mothers sometimes tell stories about how they experience themselves as being 'a rope in a tug of war' because of conflicting messages from social narratives relating to different models of childbirth care:

> Everything should be so natural, just water and relaxation and everything should go well, I thought about if I could do it. Should I try birth without analgesia? Should I try to become one of the elite groups? ... Then I got more confused, the women from the Homebirth Association and the midwife who was afraid of water [births] had a debate in the newspapers about the welfare of women during birth and they accused each other of high handedness and of being ignorant. I felt like a rope in a tug of war. I gave up; I felt that I did not have the knowledge to make a decision about what was best for me, not without help. All the messages I got [narratives] were conflicting and inconsistent. I did not trust myself to think more about it, I waited for what would happen and hoped for the best. (quoted in Ingadottir et al. 2002: 353)

In such circumstances, the care 'waited for' is likely to accord with the dominant, medical, model.

Much has also been written about how professionals learn through reflecting on their own practice (Schön 1987; Benner 1984; Jarvis 1999; Johns 2002), an action which can itself be looked at as a form of storytelling. Here too the dominant model can have an influence and it is difficult to avoid a degree of self-justification when retelling one's own story (Kirkham 1997b). Storytellers, or listeners with clinical experience, recognize clues or clinical patterns. The above narrative highlights the need of being aware of the conflicting messages women get both from health professionals and society and how midwives should find new ways of talking about childbirth with a pregnant woman based on her background and experiences.

Not all knowledge is easily visible in narrative form or as discourse (Jarvis 1999). Intuitive knowledge or subjective knowing (Belenky et al. 1997) is a kind of embedded knowledge and is different to articulated knowledge (Wengraf 2001). Even though now-forgotten midwifery skills and practical knowledge about childbirth can never be recovered, there are ways to reveal partial as well as situated knowledge, one of them being using narrative methodology where narrative knowledge (Lyotard 1984) stands as the basis of human experience and society.

Alternatively, particular narratives can contain descriptions of intuitive knowledge and ways to help the mother during pregnancy and childbirth. The narrative can be recognizable to midwives and contain elements that could be a guide for midwifery practice, not necessarily in an explicit or concrete way but as a stimulus to thought.

## Narrative in Practice: Knowledge, Truth and Certainty

Storytelling is a creative process in which the narrator and reality is reborn (Frid, Öhlen and Bergstöm 2000), thus making every story a new one depending on when and to whom it is told and what is remembered and felt to be important at that time. If the plot shifts, the story could even have a new ending. Keeping in mind Mishler's point that 'all factors of memories, motive, and context influence what they [the tellers] include in their accounts or stories, which are necessarily and irremediably selective and incomplete' (1995: 96), it is challenging to use birth stories in order to identify knowledge about childbirth and midwifery epistemologies. One could question if this is a good way of doing research as stories are always changing, but knowledge and truth are not static either.

The truth of the story told need not be the primary issue, for one can instead focus on the meaning the storyteller conveys, their explanatory conclusions and evaluations. This is demonstrated when midwives use parables in teaching, hybrid stories distilled from women's common experiences and containing details with which the present listeners can identify. Stories of the same events can also be told from different points of view or models of care, as is shown when all those present at a birth write their story of that birth.

From detailed stories, the hearers take what they find to be valuable (Riessman 1993). This makes them a democratic means of conveying information. Jordan describes how, among Maya women of Yucatan, stories function in decision making during labour:

> As difficulties of one kind or another develop, stories of similar cases are offered up by the attendants, all of whom, it should be remembered, are themselves experienced birth-givers sharing a collective expertise. In the way in which these stories are treated – elaborated, ignored, taken up as themes, characterized as typical, and so on – the collaborative work of deciding on the present case is done. (Jordan 1993: 195)

This may seem a long way removed from today's evidence-based practice, but this process can still be seen working in forums where clinical issues are debated. In Western societies, unlike that of the Maya, such gatherings do not include childbearing women.

Midwives tell stories when performing consultations on the labour ward or by telephone with other midwives and midwifery students, which is in accordance with what Jordan (1993) wrote. These stories contain situated knowledge, which is relevant to the midwives because it is pragmatic and is summoned forth as the characteristics of the situation require. It is therefore important to be aware of the context of a story, the values implicit within it and why it is being told to a particular audience.

Despite rhetoric concerning holistic and woman-centred care, there is great pressure upon midwives to give the 'right' care and ensure that women make the 'right' decisions, sometimes at the prompting of midwives themselves, as one them admits:

> [midwives] always come back to 'this is what we normally do' and women will just comply with it. If you look at Vitamin K, you've got to give the information and sometimes we slant it in a way that they would comply with us. We're very powerful and very able to do that. Not always, but when it suits us. When it doesn't, we manipulate the situation ... we give them informed choice so long as they make the choice we want and we still hold the power. I think obstetricians have had a lot of power and I think midwives are taking that power. (quoted in Kirkham and Stapleton 2001: 142)

There are honourable exceptions to the pressure on midwives to gain informed compliance but, overall, in medicalized settings working on an industrial model, midwives tend to 'go with the flow' of the organization (Kirkham and Stapleton 2001; Ólafsdottir 2006).

Analysis of midwives' stories shows how we seek to achieve professional aims by thinking in dualistic terms of 'good' and 'bad' (Fielder et al. 2004). This often involves separating ourselves from 'others'. These others may be those seen as undeserving of professional midwifery status when midwives struggle for legal recognition (Heagarty 1996; Lay 2000), those who midwives perceive as working differently from themselves (Curtis, Ball and Kirkham 2003) or clients who midwives stereotype as undeserving (Kirkham et al. 2002). Such dualistic thinking is common where there is clear expert authority. It is not surprising in a clinical environment where clinical guideless and protocols rapidly ossify into rules. Such an environment can encourage midwives to go on being passive learners as they have long been dependent on authorities to hand down the truth (Belenky et al. 1997). Yet awareness of one's own patterns of thinking, so evident in how people construct their own stories, would give midwives choices to behave in ways more consistent with the inclusive values of midwifery. Dualistic thinking means that everything between the opposing poles can be ignored. Midwives could choose to hear different valid viewpoints in the stories around them and foster them, thus developing new knowledge. Midwifery students might then learn about a diversity of opinions and dualisms might give way to multiplicity and a holistic view.

Downe and McCourt (2008) have argued that instead of taking a linear approach to evidence, we should look at science as a paradigm of an ongoing dialogue, putting forward questions and looking for answers, where decisions must be made in relative uncertainty. This dialogue can be looked at as one form of storytelling in midwifery practice and in interdisciplinary research about childbirth.

Gilligan has described how women value a web of relationships (Gilligan 1982). This can be related to the close relationship midwives and mothers have had historically and which they greatly value. Midwives stress the importance of the relationship between midwives and women, which is considered to be special or unique, and fundamental to their practice. Birth stories usually describe the woman's web of immediate relationships, and often that of the midwife too (see below). Lay described how direct-entry U.S. midwives 'offered birth stories as evidence, and a speaker's credibility was established not only by the number of births she had attended but also by her own mothering experience – her embedded knowledge' (2000: 21). A story's movement around a web of relationships and similar movements in time, which may be cyclical, seasonal or concerned with other births, also creates a web, rather than the linear shape of clinical records such as a partogram, documenting time, the hours a birth takes.

## Women's Ways of Knowing: From Oral Tales to Authoritative Knowledge?

Through historical time, birth stories present complex social and experiential knowledge. Midwives around the world have written diaries and books to document their work and to inform other midwives, students and prospective mothers (e.g., Björnsdottir 1929; Gaskin 1975; Armstrong and Feldman 1986; O'Connor 1995; Olphen-Fehr 1998). Such knowledge has not been presented or worked on in a scientific way. It is not authoritative in that it lacks power relative to the simplified and currently dominant narratives of birth, that is as something to be feared and rendered safe by experts and technology.

Based on her anthropological research, Jordan (1997) wrote about the notion of authoritative knowledge and claimed that in any particular domain several knowledge systems coexist but some carry more weight than others, either because they explain the state of things better for the purposes at hand or because they are associated with a stronger power base, usually both. Therefore one could say that the knowledge system of midwifery, which is a women's profession and less powerful than the male orientated mainstream knowledge of medicine, needs emancipation.

Studies based on midwives' stories (Crabtree 2004) support the notion that many people have lost contact with the 'natural' aspects of childbirth, as a rite of passage or life process, and that many women have lost faith in their ability to give birth in a normal way. Now that the caesarean rate has reached between 18 and 23 per cent in a variety of Western countries such as Iceland and the U.K., and only about 40 to 60 per cent of childbearing women experience normal birth (defined as vaginal birth without induction and instrumental assistance with or without anaesthesia),[1] it is not surprising that birth is no longer seen as something that women can do. It can be presumed that women and midwives have lost, along the way, some of the gifts, insights and skills of traditional midwifery, which used to be an integrated part of their culture and everyday life. In a consumer society, it is logical that many stories concern positive experience of births carried out by experts. Take the following example, from a mother speaking about her birth experience:

> When the doctor had given me the epidural, everything was fine. I felt comfortable and helped my husband with the word puzzle while we waited for the baby to come. I felt a little bit of pain, but it was not unbearable in any way ... 'It is a boy', someone said and passed over my head a slimy and bloody package. The midwife took him before I could see him ... she brought him back and put him on my shoulder, there was a tent over my breasts and my hands and arms had tubes in them so I had not much opportunity to hold him. He seemed to be a handsome boy, wrinkly and completely bald. I was so happy and I saw everything in pink colours. It is impossible to describe how overwhelmed I was with wellbeing and joy. (quoted in Ingadottir et al. 2002: 357)

As have a number of feminist writers (e.g. Ribbens and Edwards 1998; Hesse-Biber and Yaiser 2004), Belenky et al. (1997) believe that conceptions of knowledge and 'universal truths' that are accepted and articulated today have been shaped throughout history by the male-dominated majority culture. By examining basic assumptions of older academic disciplines, such as medicine, new conclusions could be drawn. New academic disciplines such as midwifery can develop their own body of knowledge by looking at birth stories, which provide a context for understanding what midwifery is. This understanding could be used as a step to identifying epistemologies that underpin midwifery practice and childbirth care and produce authoritative knowledge that has the ability to influence health care and social activity around childbirth.

## Narrative Time and the Culture of Childbirth

Culture is generated by shared storytelling in which beliefs, identities and relationships are linked to the narrative, as well as by actions and events. In this sense the culture of midwives and their identity is constructed and reinforced in their stories. Traditionally, midwives learnt their art with their mothers and grandmothers, and were in intimate and direct contact with the women they attended. Midwifery was looked upon as being a natural aspect of a woman's life. This kind of knowledge was not viewed as distinct from knowledge of other aspects of life (Donnison 1988; Page 2003; Wickham 2004). In a way it was life itself, though in many societies birth stories were only told amongst women who had themselves given birth.

Many stories told by midwives rejoice in such expressions as 'it went splendidly well' which refers to the fundamental ideology of midwifery, the belief in normality of birth. The following story is an archetypal narrative, a collective 'store of wisdom' related by a midwife who does not just construct a narrative but constructs a social world of midwifery in Iceland in 1954, though the story is not only of relevance to that time as we can also look at it from the perspective of the present. A similar story with the same atmosphere could have been told in other places all over the world but this is the only narrative written down fifty years later – it has its own life, its own perspectives, its own insights. As Riessman (1993) points out, stories shift according to the time they are written or read. Elsa, a midwife with fifty years' experience, tells how things in the mid 1950s were different, and she sees that there are aspects of the birth which would be seen as significant now but which at the time she did not think of as particularly unusual.[2]

> There was one birth on a farm. I want to tell about how it went. It went splendidly well. But when I look back now there are aspects which I did not find particular at that time. This woman was forty-three years old; she was not much younger than my mother and her eldest child was the same age as I, and I had been a babysitter for her other children when I was eight to eleven years old. So this woman was for me more like a mother than a woman giving birth. She was in her ninth pregnancy and was forty-three years old. This would have been considered abnormal [today] or at least there would have been a fuss about this birth. It was summer, a light night and it was so quiet. And there were many children. The eldest had left home. I went to her in the evening and she gave birth to a lovely, beautiful daughter. I remember that her husband was with her or he was within reach. The fathers were with their women at that time and when my mamma had her babies my dad was always with her. When the baby was born and I was ministering to her, the father went out to the kitchen. When I came out he was fast asleep on a chair with sheep clippers in his hand. This was so peaceful and looking back I think, the woman was forty-three years old and having her ninth child and the birth was wonderful. I was well into my pregnancy at the time and my first child is the same age as this one.

This narrative is structured around the place of birth and a midwifery relationship with the woman and her family. The story sheds light on the changes in the culture of birth in Iceland. Births that are now labelled as high risk or abnormal were, fifty or so years ago, 'normal' home births (out in the country) and peaceful family events in which birth was a rite of passage for the childbearing woman as well as the midwife herself.

In this narrative there are many embedded stories or narratives which cast light upon practices then (1954) and earlier, and these have relevance to the culture of childbirth and midwifery knowledge in our societies today. There are embedded stories not told about fathers being with their women before the time of hospital birth and there are stories about the environment of birth, the light summer nights. Incidentally, the word for midwife in Icelandic *ljósmóðir* means 'mother of light' and many birth stories in Iceland include narratives about the light as the baby was born. The storyteller travelled between time periods, there was also the time factor related to the age of the woman, and rules about the woman's age and criteria of high-risk pregnancy and where women are supposed to give birth – 'This would have been considered abnormal [today] or at least there would have been a fuss about this birth'.

There are many sorts of time within a story and how time is seen˚ is controlled by the storyteller. In the story above there is a doubling back to the present and there are repeated links between the mother's family time and that of the midwife – 'I was well into my pregnancy at the time and my first child is the same age'. Time, as spun in this story, is not linear but more like a web, a point which echoes Carol Gilligan's (1982) work on how women view relationships. Such a web of time and relationship seems more inclusive and supportive than linear time. Here is another more recent narrative, told by a midwife in the U.K.

> I looked after this woman who really wanted a normal birth. She had had three Caesarean sections and wanted to labour and have the maximum chance of a normal birth with her last baby. She could not get that in hospital, the rules around trial of scar [the time allowed for a woman who has a surgical, uterine scar to labour] were too tight for her, and she was not 'allowed' an NHS home birth. I booked her for a home birth as an independent midwife. We talked a lot in pregnancy and she bloomed. The supervisor of midwives was really supportive, I was so glad of that.

> She hired a pool and felt ready for the birth. She went into labour on a lovely summer's evening. She laboured really rapidly, though she hadn't really laboured before, and delivered only half an hour after I got to the house. She didn't even get into the pool. The children came in straight after the birth and admired their brother. It was a lovely family scene. She bled after that and had to go into hospital for a transfusion. But she still felt triumphant. I was a bit nervous as to how we would be received in the hospital, but most of the midwives there were fine with us. She went home the next day. I've heard her tell her story to friends and the PPH [post-partum haemorrhage] and time in hospital doesn't feature. She got what she wanted and was happy with it. I'm glad I stuck my neck out for her.

This is also a story of a home birth within a family, this time in 2004 in England. With its awareness of place of birth, light, season and family context, it is in some ways similar to the first story. From the midwife's viewpoint there is also an awareness of 'the system': in booking this woman for a home birth she is doing something which a midwife within the NHS system could not do, and the power of good or bad relationships with her supervisor and the hospital midwives and obstetricians all within the system, remain very important. A vaginal birth after Caesarean section at home is possibly a special case and this midwife cared for the woman at home only because she wanted to meet the woman's wishes, which would not have been granted in hospital.

All midwives in the Western world are now very aware of what is required of them by policies, protocols and clinical guidelines. So the awareness of these 'rules' appears in most contemporary midwives' stories, as in this fragment:

> Our system is like this, we have a lot of devil's protocols that flex around
> our feet and hinder us, but we have to follow the rules and the women do
> not fit in with the rules. Before the technology, you trusted your clinical
> skills, had this in yourself, but not in some machines.

These 'rules' often concern the time a woman is 'allowed' to be in labour,
or the time allocated for part of the labour. The woman whose story is
told above was anxious about the time allowed for the 'trial of scar' if she
was to labour in hospital. When she was able to free herself from such
constraints of timing, she relaxed and laboured quickly. This situation is
not unusual. This is not to say that labouring with a scar does not carry
risks; but the anxiety and inhibition, which result from time limits, can
also have negative outcomes. Midwives' awareness of rules and of time
limits has a real impact upon their confidence and their practice. This
awareness forms part of contemporary narratives of practice, a discourse
with authority which is not present in earlier accounts, where the midwife
often features as a solo professional in the context of a family or in an
empowering relationship with the woman herself.

Mothers' birth stories are carried with them throughout their lives.
They carry much of the cultural context of birth and can be most useful
in the education of midwives. In reading mothers' stories, which were
written or recorded in interviews in recent years (e.g. Audit Commission
1996; Kirkham and Stapleton 2001; Edwards 2005), there are various
striking cultural themes. Reference is often made to what is 'allowed'.
This may be in linear time, such as length of labour or time between feeds
for a baby. Sometimes it is what is socially allowed, such as whether it is
'all right to ring the bell' for attention in hospital. Awareness of how busy
midwives are underpins much anxiety about what is appropriate to ask as
a patient. 'If I'd have asked she might have said, but you don't like to ask'
is a theme which recurs in so many women's accounts of their maternity
care and acknowledges their awareness of how busy midwives are (KCW
1980; Audit Commission 1996; Stapleton et al. 2002).

The following birth story, which comes from a busy hospital labour
ward in the summer of 2005, shows that there is not much time for
attention and communication with the prospective parents. The story
is related by a midwifery student starting her midwifery studies at the
University of Iceland and is part of an assignment about birth stories in
her surroundings. For this, students are required to bring a birth story
to discuss in class, and then they are asked to consider the underlying
models of care and relationships with women that are present within
the story. Finally, they are asked what they can learn from it on their
journey to becoming a midwife. The story relates events that occurred to
the teller's friend, and was repeatedly told in a group of young women
sharing their birth experiences. It is noteworthy how time is carefully
reported and how the sense of 'calm' in the story is different from that in
the story from a farm in 1954:

About 7 o'clock, Anna is on the labour ward and the work there is crazy. They [the midwives] were madly running around. Anna was five centimetres dilated and they were there for about two hours, and once in awhile someone checked on them, the midwife told them that she was also delivering twins. In two hours time they were asked to go out and wait in the fathers' room [i.e., the corridor in front of the lift]. Anna was just coping and found it hard to walk because of the pain ... about 10 o'clock she gets a labour room, the midwife was in and out of the room and told them without them asking that she had been working in the morning, that she was on an extra shift now and that she had a morning shift tomorrow. Most of the time Anna and her husband were alone. There was an exchange of shifts and another fresh, lovely midwife came [but she] often had to go out. In the end Anna's husband put his foot down and said someone had to be there now, unless he was to receive [i.e., deliver] the baby, he did not trust himself to be there alone with her. The midwife stayed but instead traffic increased into the room, people opening the door asking for things... This disturbed Anna a lot... She liked her midwife and found her encouraging and liked the fact that she called her by her name. At 1 o'clock a healthy girl was born, but Anna 'would have died' if this had been her first born, as she said herself.

Midwives also get frustrated because they cannot spend time with the women they are looking after and this has an influence on their relationship with them. Here, Sara, a midwife of twenty years, comments on the problem.

I am so tired of this intrusion on the ward, the phone is always ringing outside [the labour room], this interruption, we never get peace ... you are doing many things at the same time and you don't get the time you need with the woman, there is always this interfering which can be very disturbing because you are always cutting on this connection [with the woman], always just dropping out, always leaving. This is disturbing, this interferes the birth unbelievably ... It is a deluxe if you are with one woman and then this looseness comes and you start to change your methods of working because you have to get used to this. You do less of the 'sitting over' (*yfirseta*) by the women and it has the effect that you move back, and you seek to leave [the woman] more often. I think that when there is less of 'the sitting over' it disturbs the process of birth and it of course increases the intervention rate, I think so. You do not know about the woman and she does not know about you.

The following narrative, from a midwife working in a busy central hospital influenced by the technocratic model of birth (Davis-Floyd 2003), highlights how, by necessity, midwives have to work 'with institution' instead of 'with woman' and experience conflicts between models of care (Hunter 2004; Ólafsdóttir 2006). Midwives experience stress and burnout, and the emotion work of midwives is strongly influenced by the context of their practice (Hunter 2005). As a result, midwives need support (Kirkham and Stapleton 2000; Deery 2005), as this extract from a midwife's diary entry from 2004 makes clear.

I went home in the morning without having had anything to eat or gone to the toilet ... There were thirteen women on the ward during the night, nine of them came in and five of those were high risk... We were four midwives and a first year student and three midwives were called in extra at different times.

All midwives were occupied in the labour rooms and I had three women and I ran between them .... [A]ll women had some complications and had to be looked after constantly. I felt all the time everything was going out of boundaries and that we were working in unsafe surroundings and I did not have overview of things.

When I came home I sat down at the kitchen table and took the newspaper. There my husband found me crying. I slept badly, and thought about getting another job ... this was one small example and not the only one [the third busy shift during that weekend]. It is very seldom that I go home and think 'today I worked as a midwife ... the work I love gives me pain'. When the situation is like this our midwife heart is not ticking.

When the context of birth has such an impact upon midwives, it is scarcely surprising that women are often portrayed as passive in modern dramas concerning birth. Stories, of procedures done to women or doctors rescuing mothers and delivering babies who would otherwise have suffered dire consequences, can construct the mother's role as that of someone who is rescued by the heroic doctor (or midwife). Mothers' stories also teach other women what is possible or to be expected. Women rarely expect services or practices that are not the experience of their peers, or which their local service cannot provide (Porter and Macintyre 1984), though there is a growing body of women who seek to avoid a previous negative experience. Where a service makes new options available, the reality of those options is revealed to the community through the experiences of those using the enhanced service. Where home birth is made available or peer support for breastfeeding makes breastfeeding experiences more widely known, other women's expectations change. Knowing women like themselves, whose story includes a homebirth or breastfeeding, alters women's expectations of what they might do or experience. Hearing a range of different stories may have more reality for women faced with choices than an abstract menu of options, such as those portrayed in clinical guidelines.

## The Web of a Birth Story Across Twenty Years

We conclude with the following story, one woman's experience of differences in practice and relationships between the years 1978 and 1998 in Iceland. We think it is appropriate here because we recognize the earlier story from our own experience and choose to use such stories in our teaching. This woman's story speaks for itself.

I fell into the old relationship with the patriarchy, like it had been in the year 1978, when I came [to the prenatal clinic at the hospital] there I see a doctor who knows everything much better than I, he was always telling what was best for me, I felt that he talked over my head... he was trustworthy and he wanted to do everything for me, I followed him like a dog on a leash. He didn't say a word but looked to check if I did not follow him [on the way to be examined] then he said he was going to keep my journal. I had later to steal my journal and go back to the community centre.

My situation was related to sickness and old ghosts from before appeared, - felt I was losing control. But it was great support to be in contact with the same persons [a midwife and a midwifery student]. To be in a stable relationship with the same persons who knew my expectations. To have personal relationship through the pregnancy with people I know and have them with me during the birth is invaluable. There is a connection that is difficult to describe with words some kind of empathy, caring and complete trust.

It is not possible to compare this; then I felt that I had been invaded, in the most private part of myself. It is maybe rude to say it, but this was next to, being raped. I was there wide open, and there were people running in and out, I felt dirty. I felt I was alone, very much alone, yet the room was full of people. It would have been good to be able talk about this experience afterwards ... this negative experience followed me for many years. At that time the baby was just a health problem, as he was taken with a vacuum extraction, and he was taken away. This time I was involved all the time, not under the influence of drugs. I felt when he moved his fingers during the birth, and that sense made mother's love flow out of my ears. I got him in my arms, new, wet, and he smelled so good and I felt this connection, breastfeeding and so on...

It is difficult to describe how it is to go through labour without getting any analgesia. I felt as I emerged with a universal mother. My grandmother and my great grandmother were both capable to do this and so could I. I saw myself as a strong woman. I would be able to give birth just as women were able to do through centuries. I suddenly got this feeling of being a woman. I felt like a universal woman! After my first childbirth in 1978, my only thought had been to crawl back into my bed; underneath the sheet hoping that I would never have to see these people again. I had no self-respect left and I felt as I was just a crotch. I did not have a name, I felt just like a crotch having a baby. This time I felt like a universal woman. (quoted in Ingadottir et al. 2002: 55–6)

While this story shows how maternity services can be abusive and disempowering, it demonstrates that progress has been made in maternity care. The above narrative reflects the changes in services and ideology in midwifery models of care as they are perceived by one woman. These changes are visible in the ideology of midwifery education at the University of Iceland (Olafsdottir 1995) and in publications such as the 'Changing Childbirth' report in the U.K. (DoH 1993), which called for women-centred services to meet individual needs, moving from patriarchy to informed choice, control and continuity of care. These concepts have had rhetorical influence in the development of models of midwifery care, and they are related to common conceptualizations of midwifery care in Western countries. This story shows the way a narrative can be used to interpret previous experience and link with the experiences of others, between countries and the universe. Such a network of links, made by a story, enhances the lives of individuals and communities and justifies the efforts of midwives to support women.

## Notes

1. For statistics on the situation in Iceland, see: Georsson et al. (2006). Figures for Britain are taken from those published by the Department of Health's Statistical Service: see, 'NHS Maternity Statistics, England: 2005–2006'. Retrieved 1st December 2008 from: www.ic.nhs.uk/pubs/maternity0506.
2. The interview transcripts reproduced in this section are taken from Ólafsdóttir (2006).

## References

Armstrong, P. and S. Feldman. 1986. *A Midwife's Story*. New York: Ivy Books.

Audit Commission. 1996. *Maternity Care Survey*. Bristol: Audit Commission and Institute of Child Health.

Belenky M.F. et al. 1997. *Women's Ways of Knowing*. New York: Basic Books.

Benner, P. 1984. *From Novice to Expert: Excellence and Power in Clinical Practice*. Menlo Park, CA: Addison Wesley.

Björnsdottir, T.A. 1929. *Nokkrar sjúkrasögur úr fæðingabók* [*A Few Cases from a Birth Book*]. Reykjavík: Herbertsprent.

Crabtree, S. 2004. 'Midwives Constructing "Normal Birth"', in S. Downe (ed.) *Normal Childbirth: Evidence and Debate*. Edinburgh: Churchill Livingstone, pp. 85–99.

Curtis, P. , L. Ball and M. Kirkham. 2003. *Why Do Midwives Leave? Talking to Managers*. London: Royal College of Midwives.

Davis-Floyd, R. 2003. *Birth as an American Rite of Passage*, 2nd edn. Berkeley: University of California Press.

Deery, R. 2005. 'An Action-research Study Exploring Midwives' Support Needs and the Effect of Group Clinical Supervision', *Midwifery* 21: 161–76.

DoH. 1993. *Changing Childbirth. The Report of the Expert Maternity Group*. London: HMSO for the Department of Health.

Donnison, J. 1988. *Midwives and Medical Men: A History of the Struggle for the Control of Childbirth*. London: Historical Publications.

Downe, S. and C. McCourt. 2008. 'From Being to Becoming: Reconstructing Childbirth Knowledge', in S. Downe (ed.) *Normal Childbirth: Evidence and Debate*, 2nd edn. Edinburgh: Churchill Livingstone, pp. 3–24.

Edwards, N.P. 2005. *Birthing Autonomy*. London: Routledge.

Fielder, A. et al. 2004. 'Trapped by Thinking in Opposites?' *Midwifery Matters* 102: 6–9.

Frank, A.W. 2000. 'The Standpoint of Storyteller', *Qualitative Health Research* 10(3): 354–65.

Frid, I., J. Öhlen and I. Bergbom. 2000. 'On the Use of Narratives in Nursing Research', *Journal of Advanced Nursing* 32(3): 695–703.

Gaskin, I.M. 1975. *Spiritual Midwifery.* The Farm, Summertown TN: The Book Publishing Co.

Georsson, R.T. et al. (eds), 2006. *Fæðingaskráning, 2006. Skýrsla frá faedingarskráningu* [*National Documentation of Births in Iceland, 2006. A Report of Births*]. Reykjavik: Women's Department and Children's Clinic, University Hospital of Reykjavik.

Gilligan, C. 1982. *In a Different Voice: Psychological Theory and Women's Development.* Cambridge, MA: Harvard University Press.

Greenhalgh, T. and B. Hurwitz. 1998. *Narrative Based Medicine: Dialogue and Discourse in Clinical Practice.* London: BMJ Books.

Heagarty, B.V. 1996. 'Reassessing the Guilty: The Midwives Act and the Control of English Midwives in the Early Twentieth Century', in M. Kirkham (ed.) *Supervision of Midwives.* Hale: Books for Midwives, pp. 13–27.

Hunter, B. 2004. 'Conflicting Ideologies as a Source of Emotion Work in Midwifery', *Midwifery* 20: 261–72.

——— 2005. 'Emotion Work and Boundary Maintenance in Hospital-Based Midwifery', *Midwifery* 21: 253–66.

Hesse-Biber, S.N. and Y.L. Yaiser. 2004. *Feminist Perspectives on Social Research.* New York: Oxford University Press.

Hulst, L. and E.R. van Teijlingen. 2001. 'Telling Stories of Midwives', in R. Devries et al. (eds), *Birth by Design.* New York: Routledge, pp. 166–79.

Ingadottir, E. et al. 2002. *Konur með einn í útvíkkun fá ekki samúð. Fæðingasögur íslenska kvenna* [*Women with 1 cm Dilation Get No Sympathy: Birth Stories of Icelandic Women*]. Reykjavik: Forlagið.

Jarvis, P. 1999. *The Practitioner Researcher: Developing Theory from Practice.* San Francisco, CA: Jossey-Bass.

Johns, C. 2002. *Guided Reflection: Advancing Practice.* Oxford: Blackwell.

Jordan, B. 1993. *Birth in Four Cultures.* Prospect Heights, IL: Waveland.

——— 1997. 'Authoritative Knowledge and its Construction', in R.E. Davis-Floyd and C. Sargent (eds), *Childbirth and Authoritative Knowledge: Cross-Cultural Perspectives.* Berkeley: University of California Press, pp. 55–79.

KCW. 1980. *Maybe I Didn't Ask.* London: Kensington, Chelsea and Westminster Community Health Council.

Kirkham, M. 1997a. 'Stories and Childbirth' in M. Kirkham and E.R. Perkins (eds), *Reflections on Midwifery Practice.* London: Balliere Tindall, pp. 183–204.

——— 1997b. 'Reflection in Midwifery: Professional Narcissism or Seeing with Women?' *British Journal of Midwifery* 5(5): 259–62.

——— 2004. *Informed Choice in Maternity Care,* Basingstoke: Palgrave.

Kirkham, M. and H. Stapleton. 2000. 'Midwives' Support Needs as Childbirth Changes', *Journal of Advanced Nursing* 32(2): 465–72.

——— (ed.) 2001. *Informed Choice in Maternity Care: An Evaluation of Evidence Based Leaflets.* York: NHS Centre for Reviews.

Kirkham, M. et al. 2002. 'Stereotyping as a Professional Defence Mechanism', *British Journal of Midwifery* 10(9): 509–13.

Kvale, S. 1996. *Interviews: In Introduction to Qualitative Research Interviewing.* London: Sage.

Lay, M.M. 2000. *The Rhetoric of Midwifery: Gender, Knowledge and Power*. New Brunswick, NJ: Rutgers University Press.

Lyotard, J.F. 1984. *The Postmodern Condition: A Report on Knowledge*, trans. G. Bennington and B. Massumi. Manchester: Manchester University Press.

McCloskey, D.N. 1985. *The Rhetoric of Economics*. Madison: University of Wisconsin Press.

Mishler, E.G. 1995. 'Models of Narrative Analysis: A Typology', *Journal of Narrative and Life History* 5(2): 87–123.

O'Connor, M. 1995. *Birth Tides: Turning Towards Home Birth*. London: Harper Collins.

Olphen-Fehr, J. van. 1998. *Diary of a Midwife: The Power of Positive Childbearing*. Westport, CT: Bergin and Garvey.

Ólafsdóttir, O.A. 2006. 'An Icelandic Midwifery Saga – Coming to Light: 'With Woman' and Connective Ways of Knowing'. Ph.D. dissertation. London: Thames Valley University.

——— (ed.) 1995. *Curriculum of the Educational Programme in Midwifery at the University of Iceland*. Reykjavik: University of Iceland.

Page, L. 2003. 'One-to-one Midwifery: Restoring the "With Woman" Relationship in Midwifery', *Journal of Midwifery and Women's Health* 48(2): 119–25.

Polkinghorne, D.E. 1995. 'Narrative Configuration in Qualitative Analysis', in J.A. Hatch and R. Wisniewski (eds), *Life History and Narrative*. London: Falmer Press, pp. 5–24.

Porter, M. and S. Macintyre. 1984. 'What Is, Must Be Best: A Research Note on Conservative or Deferential Responses to Antenatal Care Provision', *Journal of Advanced Nursing* 19(11): 1197–200.

Ribbens, J. and R. Edwards. 1998. *Feminist Dilemmas in Qualitative Research: Public Knowledge and Private Lives*. London: Sage.

Riessman, C.K. 1993. *Narrative Analysis*. London: Sage.

Schön, D.A. 1987. *Educating the Reflective Practitioner*. London: Jossey-Bass.

Simpson, J. 2004. 'Negotiating Elective Caesarean Section: An Obstetric Team Approach', in M. Kirkham (ed.) *Informed Choice in Maternity Care*. Basingstoke: Palgrave, pp. 211–36.

Stapleton, H. et al. 2002. 'Silence and Time in Antenatal Care', *British Journal of Midwifery* 10(6): 393–96.

Wengraf, T. 2001. *Qualitative Research Interviewing*. London: Sage.

Wickham, S. 2004. 'Feminism and Ways of Knowing', in M. Steward (ed.) *Pregnancy, Birth and Maternity Care*. Edinburgh: Books for Midwives, pp. 157–68.

*CHAPTER 9*

# HOW LONG HAVE I GOT? TIME IN LABOUR: THEMES FROM WOMEN'S BIRTH STORIES

*Christine McCourt*

She [the midwife] ought to be patient and pleasant; soft, meek, and mild in her temper, in order to encourage and comfort the labouring woman. She should pass by and forgive her small failings, and peevish faults, instructing her gently when she does or says amiss: But if she will not follow advice, and necessity require, the midwife ought to reprimand and put her smartly in mind of her duty; yet always in such a manner, however, as to encourage her with the hopes of a happy and speedy delivery
—John Maubray, *The Female Physician* (1724)

A labour which is unduly prolonged is likely to give rise to one or more of three types of distress, namely maternal, foetal or 'obstetricians' distress'. Of the three, the last may be easily the most dangerous!
—Donald (1969)

The role of the attendant in labour, whether midwife or doctor, is to measure, record, interpret and then treat.
—Stallworthy and Bourne (1979)

## Introduction

This chapter draws on women's narrative accounts of their birth experience in a London hospital to look at the relationship between how time is managed in a modern maternity unit and how it is experienced by women in labour. These narrative interviews were conducted as part of a larger study evaluating attempted reforms in the U.K. towards a more 'woman-centred' maternity service.

We discussed in Chapter 3 (this volume) the active management of labour in obstetrics, and how time is mapped out and managed in this system.[1] The partogram used in this approach gives a linear, graphical representation of time and the progress of labour. Under an active management system, action lines are used to trigger intervention when the progress of labour, measured in terms of cervical dilatation against time, does not reach a certain pace, and norms of progress and time are used as a standard against which all labours should be measured. Both the partogram and the active management approach convey the idea that, through documenting and managing time, control over the complex, physiological process can be assured. This idea remains very resonant despite the lack of scientific evidence to support it, suggesting that it may have an important ritual role when birth takes place in a hospital setting.

This chapter focuses particularly on the themes in the narratives of the women interviewed which relate to time, and which emerged as being quite different from the assumptions inherent in the active management approach of modern obstetrics. This difference in senses of time, which was clearly a source of discomfort to the women, is reflected in the tenor of their stories. The chapter looks at the women's embodied experiences of time in labour and considers these in their context; it also considers the nature of the obstetric unit where they gave birth and the wider culture of medicine, healthcare and ways of seeing and doing which influenced the environment in which they gave birth.

## Background and Methods

As part of a large study of reforms in the organization of maternity care, which looked at a number of clinical and organizational parameters, we interviewed a sample of women about their experiences of care in the two systems we were evaluating (McCourt 1998). In this study a narrative interview approach was used in order to overcome some of the limitations of other commonly used methods in health services research. The sample of women was taken from a larger survey, where women had been asked to complete a series of postal questionnaires – once late in pregnancy, then again two weeks and then thirteen weeks after birth – on the care that they had received and their views on it. While this was useful, especially for providing a large-scale set of comparable responses, which could be summarized with statistics, we felt a survey of this kind also had important limitations. First, there was the problem of reaching and hearing the voices of women who do not respond to questionnaires. The second problem was that questionnaires, no matter how carefully designed, are based on what the researchers see as matters of interest or importance. The questionnaire design is necessarily fixed and reductive. It does not cope well with ambiguity or complexity, and cannot allow respondents to frame the issues in their own ways or words. Even though we asked women to give open comments at several points in the questionnaires, we felt that it was important for our understanding to collect accounts that were not framed in this way. We also felt it was important to include women's perspectives, as well as those of professionals, to provide a counterpoint to the accounts of labour and birth that are produced by official records, such as medical notes, and often relied upon for research evidence (McCourt 2006).

The study took place in a NHS Trust in the U.K. with two maternity units, both based in teaching hospitals. Both were 'consultant-led units', as is the case with most hospital maternity care in the U.K., although their teaching-hospital status meant a higher than average proportion of medical students and a high level of medical staffing. The larger unit had about 4,000 births per year and was known for its medical approach, while the smaller unit had about 1,000 births per year, and was thought by professionals to be more midwifery-oriented. However, both units could provide specialized care to women referred for medical problems, as well as general care for local women. The small unit was located in an ethnically and socially diverse area, with high levels of social deprivation in surrounding housing estates, while the larger catered for a more middle-class, though still diverse, local population.

In the U.K. in the 1990s, there was a policy impetus to return toward a more women-centred way of providing care. This was influenced by decades of criticism of the medicalization of care by women's, childbirth and midwives' groups. This was itself part of a longer history of medicalization in the twentieth century, with ever-wider areas of life being seen as the domain of medicine, and increasingly based in hospitals (Lewis 1990; and see Chapter 1, this volume). Home births had changed during the century from being the norm to being something which happened only in a minority of cases (about 1 per cent in 1990, rising to about 2 per cent over the next decade, following policy changes that attempted to reintroduce the option of home birth). Also, there had been a shift towards more centralized healthcare, in larger units, with the aim of delivering economies of scale and concentration of facilities and expertise. Small units with 1,000 births or less per year were being progressively closed, although in some areas, midwife-led units (often called Birth Centres) were opened at the beginning of the twenty-first century in response to consumer and midwife resistance to these centralizing trends (see Chapter 6, this volume).

The women interviewed were, therefore, giving birth in a unit with high levels of medical staffing and a medicalized ethos, where midwives did not usually practice with a high degree of autonomy. The scheme being evaluated – caseload midwifery – was intended to restore some of the characteristics of traditional midwifery, as had been practised in the U.K. before the advent of the NHS's modernist 'beds for all' policy (Allison 1996). This research is described in more detail in Chapter 5 (this volume), which discussed the very different organizational models and approaches to managing time the two groups of midwives operated with within the same overall health service. It enabled greater continuity of care for the women and greater autonomy of practice for the midwives. These contrasts were reflected in the themes of the women's narratives, in particular, the disjuncture between the embodied experience of labour and the institutional system of conventional maternity care, which was not found for women receiving caseload care.

## Themes in the Women's Narratives

A number of themes emerged from the women's narratives relating to both concepts and management of time, and these will be discussed in turn. The first theme – when does labour begin? – illuminated the disjuncture between women's embodied experiences and the authoritative knowledge of when labour begins properly. This is sometimes referred to among maternity professionals as the 'latent phase' of labour versus the 'active phase', but the idea of a latent phase is rarely conveyed to birthing women, leading to confused messages about how long labour takes. It also illuminates the importance in the authoritative discourse and practice of obstetrics of measuring and managing time in labour, which requires a clear beginning and end rather than more gradual and indeterminate shifts.

The following, linked themes – how long does labour last? and, when does labour end? – illuminated the disjuncture further. From the women's narratives, it was clear that the pattern of labour was important to women, and should not be too rapid (leading to fears about the intensity of pain or the changes taking place) or too slow (leading to exhaustion). It should also follow a wavelike pattern (found in normal, physiological labour), with a gradual build up of the length and intensity of contractions and a reduction in the time interval between them. Labour that was induced or speeded up did not follow this rhythmic pattern and was experienced as more painful, difficult and frightening. Obstetric definitions of labour are generally shorter than perceived by women, and these norms are used to guide interventions to speed the onset and progress of labour, even though (as discussed in previous chapters) they are not based on good scientific evidence and the norms have shifted historically with changing attitudes and practices.

A third theme was that of transition. While the focus in biomedical care is on the short-term physiological transition of labour and delivery, for women the physiological time frame is longer. It is also situated within the life-transition of giving birth, so that signs of labour and birth are imbued with many layers of meaning.

### When Does Labour Begin?

Time was written into the interviews – quite unselfconsciously, since we had not realized that time would emerge as such as strong thread in the analysis – in the opening question: tell us about what you did when you thought labour had started. This, of course, invited the women to consider when or what that point in time might be, and to describe it in their own way, and perhaps in addition it implied a sense of definite beginning, as is conveyed in so many medical and midwifery texts as well as in the antenatal classes given to pregnant women and their partners. Knowing when labour begins is represented as being important to women because of the way time in labour is conceptualized and managed, but also because women in industrialized countries are expected to move themselves from the home to the hospital at the key point of transition, the beginning of active labour. Knowing when labour begins, in medical terms, is therefore important to using the health services correctly.

In conventional care, women worried about getting this right because they did not want to be admitted to hospital too early or too late, and so lose face or control. Often, when these women rang the labour ward, or arrived for admission, they felt they were made to feel rather foolish and unable to trust their own bodily experiences, which they were often advised were not 'real' labour. This woman, for example, felt embarrassed as well as disappointed when she arrived at the hospital to be told she was not in 'established labour':

> I remember the midwife saying , 'well I'm afraid you're not really far along at all, in fact you've hardly dilated'. I remember thinking, 'I could have days of this' and I felt really stupid because I felt I'd come in too early. I felt I showed myself up not really knowing what was going on.

This was reinforced by the language of health professionals, who could make women feel that they were not using the service properly, if they arrived too early (or too late).

Similar issues of time and appropriateness have been discussed for other areas of healthcare (Ehrich 2003). This language echoes a range of key texts which, although they discuss the indeterminacy of the start of labour, also talk of 'false' or 'spurious' labour and emphasize the importance of determining the diagnosis, course and progress of labour precisely and correctly. Some discuss a latent phase and some do not discuss this at all. The term latent phase is used to refer to early stages of labour (before the active phase of cervical dilatation) in which the body 'prepares' for labour and the cervix softens and shortens. While linear models found in standard texts tend to treat these gradual developments as though they can be clearly demarcated and talk of 'diagnosis' of labour, many women's narratives suggest a more fluid and tentative type of transition.

In earlier chapters it has been suggested that childbirth is a major, life-changing event, which can be viewed as rite of passage. Given this, it is perhaps not so surprising that a clear beginning is sought. However, the women's accounts of early signs and symptoms conveyed the sense of labour as something which develops and gathers pace gradually, rather than following a clear linear trajectory with a definite beginning. Some talked of 'nesting' behaviours in the run-up to labour (see also Chapter

6, this volume) such as 'cleaning out the oven', shopping and even decorating, which seemed to signify an impending life change.

Most talked of embodied signs – pains or discomforts, feeling different, body fluids and leakiness. These also were mostly gradual, rather than sudden. They talked of 'twinges', or period like pains:

I remember waking up, thinking my period must have started

faint contractions ... felt like mild period pains

As already mentioned, in some cases these signs were very tentative and stopped and started over one or a few days. Early, mild, often irregular pains as described by the women are called Braxton Hicks contractions in textbooks and sometimes in antenatal classes, where they might also be described as 'practising for the real thing'. Such terms suggest the idea of a transition between states with clear boundaries, whereby such contractions occur before 'real' labour begins.

Midwives are aware of physiological theory concerning labour in which the gradual physiological changes are influenced by a very complex interplay of hormones, particularly the synthesis and uptake of oxytocin, which is important for stimulating the physiological processes of labour and breastfeeding, and of prostaglandins. The hormonal changes are known to be closely inter-related with emotional and environmental influences (Haddad, Morris and Spielberg 1985).

However, the women's accounts did not indicate that they had been told about the physiological changes involved in the latent phase and professionals' responses to their questions and their reporting of 'signs' did not seem to include any acknowledgement of this often slow and gradual prelude to active labour. Instead, this phase was usually described by professionals as 'not real' or at best 'a practice'. If the woman was in pain and could not relax, time in this early phase became an issue – time passing was being spent in a way which was not valid according to authoritative childbirth knowledge. Women described feeling physically or emotionally tired and demoralized by the time 'real' labour begins:

I was very, very tired for a long time. The labour was well, in my books, thirty-six hours, from when I started contracting to when I actually had him. The hospital doesn't say it was that long because they count established labour (laughs) but as far as I was concerned I didn't get any sleep the first night so, you know, so physically I was exhausted.

Another woman, when asked if there was anything she would wish to change, said:

Matter of fact comments by the midwife on duty, and telling me I wasn't in labour because my cervix wasn't dilated..... [I would have liked] just a bit of moral support.

It was noticeable from the women's accounts of conventional care that little explanation or support was offered to help them understand and cope with a long latent phase. As a result, midwives' preference for women to be resting or distracting themselves at home in this early period was easily and often interpreted by the women themselves as being dismissed from hospital because they were not 'in (true) labour':

> They sent my friend home because they said I was not in established labour. It would have helped me to cope in the early stages if she had been there. I was having to walk up and down the ward by myself.

However, the home environment was more likely to offer the comfort needed, if not professional reassurance.

Midwives who worked with a caseload (see Chapter 5, this volume) did not have to maintain clear boundaries, since they were assigned to women rather than wards or areas, and they attended women's homes or provided telephone support in early labour, thus their experience of time in labour mirrored the women's experiences of transition more closely. The women could call the midwife when they thought they were in labour, or if they felt they had some early signs of its imminence. This echoes the theme in Leap and Hunter's (1993) oral history of traditional midwives or handywomen as 'the woman you called'. The advice was often to relax in a warm bath and sleep, if during the evening, or during the day to be active so as to encourage labour progress and distract the women from discomfort and time passing. As a result, the women felt more comforted and less concerned about having to 'correctly' pinpoint the time when labour started properly. Those women who had female relatives able to give them this advice and support also had a different experience. This woman, for example, had her midwife call to examine her at home:

> She said it was likely to be just a practice for the real thing and that I should have a bath and go to sleep and that the contractions would probably go away, and that is just what I did, and that is just what happened.

Another woman, by contrast, had a very fast labour, but also found this approach helpful:

> I think I would have been much more frightened if I hadn't had the midwife around because it all happened so quickly and if I had been on my own just the time before getting to the hospital and would have been really quite frightening.

As a result of this experience, over time caseload midwives began to take a more flexible approach to place as well as time in labour, discussing with women as the labour progressed where they wanted to be, whether at home or in hospital. This enabled them to respond more closely to the woman and her labour, and its most appropriate time and space.

This difference in approach to managing and understanding time in early or latent labour suggested that ways of organizing care influences the ways in which professionals think about and behave with time in childbirth. The caseload midwives were not attached to ward routines, and so were able to encourage the women to respond in a more flexible way to the passing of time in labour.

As discussed in Chapter 3 (this volume), the appropriate length of pregnancy is also a matter of obstetric management, fuelling professionals' and women's concerns about accurately diagnosing the beginning of labour. If a woman's pregnancy became 'post-term' in this maternity service (defined routinely as forty-one weeks,) she would be strongly and firmly advised by obstetricians to accept an induction of labour. When a medical approach is used to induce labour, this generally commences with insertion of a gel (a synthesized form of prostaglandin) to 'ripen' the woman's cervix, so that it can open with contractions, which may be induced by means of a drip of synthesized oxytocin during labour.

Like women who had a slow and gradual early or latent phase of labour, and did not have the reassurance they needed to rest or relax at home, women who had induced labours suffered from very painful and tiring periods before 'established labour' was deemed to have commenced. Several commented on how difficult they found it to cope with this period, in hospital and isolated from their usual means of comfort and support. This woman, for example, had an induced labour and found the early phase before active labour started very difficult to cope with:

> I just wanted to be in a home environment and I'd been where I would be in the hospital and that wasn't really somewhere I wanted to spend a great deal of time when I was in labour and you only get a tiny little cubicle with just a curtain round the bed.

Another woman was upset because of being induced during the night, with the expectation that she would 'go into labour' the next morning, during usual working hours. She was left quite alone without her husband's or midwife's support since she was not considered to be in labour:

> I think they really thought nothing would happen that night, you'd just go in and they said they'd start again at six the following morning ... and the fact that something happened, well mentally I wasn't prepared for it.....
> um, it was very hard being on my own and I didn't really think about it till afterwards and that was one of the things when you did the questionnaire because actually afterwards you think, when you go to your parentcraft class, you know – sit on your chair and do this, massage your back, and all the things you expected to be doing at home with your partner – I was actually wandering around aimlessly all night on my own.

In Chapter 3, Downe and Dykes discussed the active management approach in obstetrics, which requires labour to have a definite 'diagnosed' beginning and end and demarcates the process into apparently clear and distinct phases. The latent phase was discussed above, and it was pointed out that this theoretical knowledge was not shared with women in conventional maternity care, so that they felt their early, embodied experience was invalid. One possible reason for this is that this phase is not seen as of concern to obstetrics, unless there is a perceived need to intervene to give labour a definite and timely start by induction. This is also applied to women whose 'waters break' before labour, known medically as pre-labour rupture of membranes. Such ruptures are seen as out of time, and therefore risky and requiring medical intervention – by induction – within a very limited time. The caseload midwives in this study had, however, referred to the lack of scientific evidence for this risk,[2] and negotiated a delay of up to seventy-two hours in which the woman could rest at home and prepare for the onset of contractions, with expectant monitoring by the midwife. They seemed less concerned about labour having a definite time of commencement, and more able to manage a process of gradual build up, owing to the flexible and woman- or community-centred nature of their work.

### How Long Does Labour Last?

When I first started working on this project, as a woman who had already given birth, I was very surprised by the length of time recorded for labours in medical notes. They seemed very short, at only about eight hours in many cases. I began to feel I must have been rather unusual physically. As I analysed many sets of notes I realized that many women had their labours speeded up by amniotomy ('breaking the waters') and oxytocin drip. I also noticed that most of these women were advised to have, and accepted, epidural pain relief because of the severe pain that speeding up labour can induce. The length of labour was also, in general, defined from the point of admission to the labour ward in active labour, rather than from the point in the woman's own account, which in their narratives was almost invariably longer.

The length of labour is important to women, as well as to professionals, because they experience pain and anxiety about how they will manage labour, and about the outcome. Although a woman in a Western context of good maternal health and ready access to medical intervention has less to fear about obstructed labour (Van Hollen 2003; Wilson 1995), if labour takes too long, and especially if she loses sleep, the woman can become very tired and feel less able to cope. Many women explained that their requests for epidural pain relief were due to tiredness as much as pain, as the numbing of the pain would allow them to rest, even sleep. Many felt they had been in labour for far longer than their medical notes indicated.

The women described labour contractions as wave-like in character, which implied movement and progress, so that the gradually building pain and intensity of the contractions could be experienced as positive. This wave-like pattern also helped them to cope with pain, giving them space and time to rest between contractions. When labour had been pharmacologically induced or speeded up, however, the women's contractions did not follow this physiological pattern and were more likely to be continuous or irregular, making it hard for them to cope and work with the pain rather than fight it. This woman reported of her induced labour:

I found it absolutely awful, I really did. I was like that for about five to six hours until I basically thought I can't stand this anymore, I still wasn't having normal contractions, I wasn't having any breathing space between... they were very, very long, and, yes, you can use all those breathing exercises you've learnt, and whatever, for the first two minutes, then you just basically fight the pain, you know. I felt this being unhelpful to the baby.

Because her labour was augmented by an amniotomy, another woman, delivering her third baby, began to experience continual contractions, with no relief from the pain:

If I'd known it was that painful I would have waited, even if it meant going home and coming back. I don't think it is natural and I don't think it is supposed to happen. I can't describe the pain. I think I thought I was going to die. Because I am a Christian ... I was praying because I thought this is it, I'm dying.

These differences meant that although many women feel concerned about how long labour will last they do not necessarily find medical speeding up of labour helpful or acceptable. The 'cost' to them is in terms of greater pain and fear from the intensity of pharmacologically induced or augmented contractions. While obstetric discourse suggests that objective norms of labour time can be defined, those norms have, however, shifted over time. This shift in norms leads to, or (in a dialectical process) is even produced by, greater use of interventions to speed up the progress of labour. The greater use of interventions in turn comes to shift the perception of norms of time (Albers 1999; Downe 2004; and see Chapter 3, this volume).

Llewellyn-Jones, for example, stated in the third edition of a widely used obstetric text that the first labour lasts about thirteen hours and subsequent labours about eight hours, and that as about 85 to 90 per cent of 'primigravidae' will have delivered within eighteen hours it is customary to define prolonged labour as more than eighteen hours (1982: 350). He commented that a decade before, prolonged labour had been defined as forty-eight hours. Nonetheless, by the following edition, four years' later, the norm had been revised to twelve hours and an 'aggressive active approach' recommended: 'the longest acceptable duration of labour is twelve hours from admission. If the woman has not delivered by this time, a caesarean section is performed or a forceps delivery is attempted' (1986: 128). By 1997, Miller and Hanretty's text stated simply 'the first stage of labour in a primigravida lasts up to twelve hours and sometimes longer and in a parous woman is usually 4–8 hours' (1997: 235). Stallworthy and Bourne, in another popular text, stated (discussing the importance of frequent vaginal examinations) that 'with active management of labour the information obtained from vaginal examinations is essential and no patient will be in labour long enough to have too many examinations' (1979: 143). In 1985, a BBC Continuing Education text stated that 'the accepted rate of progress is that women having a baby should reach full dilatation of the cervix by twelve hours after the onset of labour and by eight hours in a woman expecting to give birth to her second of subsequent child' (Huntingford 1985: 81), although the author then goes on to discuss the drawbacks of this theory as a definitive guideline. The title of the BBC text – 'Birth Right: The Parents' Choice' – is noteworthy since it uses consumerist language, yet the tone of mainstream texts such as the fourth edition of Llwellyn-Jones (1986) remained authoritative, with no need to use bibliographic references, although a small section in the edition introduced the concept of parental choice as a new issue for obstetricians to consider.

Labour is officially defined, in the unit studied here and in obstetric units internationally, as beginning when the woman's cervix is 4 cm dilated, also considered the commencement of active labour. Other measures of progress are also noted on the partogram – the state of the cervix and the position of the baby's head in relation to the brim of the pelvis. From the first examination, regular recordings are made of cervical dilatation and the descent of the baby's head against time. An electronic monitor is usually used to trace the baby's heart rate and the rate and strength of contractions.[3] This is the first stage. As discussed, women were routinely encouraged to accept amniotomy or augmentation with a hormone drip if the expected pace was not maintained.

The first stage is defined as ending when the woman's cervix is dilated by 10 cm – in effect when the cervix can no longer be felt around the baby's head – the woman then enters the second stage. Strict protocols were applied to the length of the second stage of labour, which should not last for longer than an hour. For women with epidural pain relief, because of its numbing effect, women were allowed an hour of 'rest' before an hour of active pushing:

> she expected me to be fully dilated at which stage they give me an hour to rest and then an hour pushing. And that's the hospital policy...

> [the midwife] emphasized that I only had an hour in which to push the baby out. After that time if I hadn't managed to produce a baby then they'd have to consider some sort of intervention. So she laid out the rules quite clearly to me.

> I pushed for about one hour and then it was getting a bit late – you know, taking too long – and there were already doctors sort of hovering around to see if I needed some help.

> and, um, I got this tenth centimetre and they let me push and they said they let you push for an hour.

> I only saw one [doctor] who came in to tell me that I only had one hour to push the baby out.

This one-hour rule, which was described by a number of women as 'like the clock ticking against them', has been adopted in an number of maternity units worldwide as a matter of custom and practice, based on professional perspectives rather than research evidence. It was presented as authoritative and in the women's interests.

The rationing of time was also reflected in the notion of time as limited and so not to be wasted, an attitude resonant of Thompson's (1967) work on the historical and social transformation of time concepts (see Chapter 1, this volume). Contractions linked time and progress. As a result, women felt that time in labour must be used and spent properly, a view conveyed clearly by some midwives:

WOMAN:          And then, when it was the second stage of actually trying to push, I kept getting cramp in my leg. It sounds a really silly thing to get but ...
RESEARCHER:  No, it doesn't surprise me.
WOMAN:          I had cramp in my leg and I was actually trying but it was very painful and the midwife, she got very angry with me.
RESEARCHER:  Did she?
WOMAN:          She sort of kept saying, 'Oh, you're wasting the contractions, you're ruining everything'. And she, you know, she was quite angry with me, saying 'Oh, you're ruining everything'. And, well, this was my special day of having a baby...

It was clear from the women's accounts that, in their experience, labour should not be too slow or too fast. If too slow, as we have seen, exhaustion set in, aggravated by both a lack of sleep, if the woman was unable to relax in early labour, and the withholding of food and drink, which was policy at the time in this maternity unit. If too slow, it was also possible to become fearful without reassurance that all was going well. If too fast, the woman did not have the time to adjust and allow herself to flow with the bodily rhythms of labour, especially if much of this time was spent travelling to the maternity unit, arriving and settling in to an unfamiliar environment.

> Looking back, I think I was doing OK. It had been a bit sudden, a bit frightening. It all started happening a bit fast. I wanted to get there to be with her [midwife]. It felt comfortable.

> I was trembling and shaky. Because it all happened so quickly I was slightly in shock. So she [midwife] suggested I have a warm bath, which I liked the idea of, just to try to relax, because I felt like I wanted to get a bit more in control as well, because it was all happening so quickly.

From the women's viewpoint, then, there was also an appropriate time and duration for labour, but its duration was counted in a different way from that of obstetrics, being more centred on their personal and embodied experience and life context, including the experience of transition from pregnancy to parenthood. A woman's experience could also involve losing their sense of time and place in labour as she focused inwards on the rhythms of her own bodily experience. As one woman put it, 'I think

I was well wrapped up in myself. Everything else around me was, like, oblivious [sic]'. The focus of the process, and its markers, differ.

### *When Does Labour End?*

For Western women for whom access to hospitals and medical facilities is the norm, there is little awareness of what professionals call the 'third stage' of labour, when the placenta is delivered. With obstetric management, the general concern to speed up processes extends to this phase, the reason being to prevent heavy blood loss after the birth. Women are normally given an injection of synthetic hormones, as the baby's head is delivered, to ensure that the placenta separates and can be delivered within minutes rather than perhaps half an hour. Many women are not aware that this takes place because of its timing, when their focus is on the baby's birth and because their consent to the procedure is rarely explicitly sought. Thus the time of the third stage is quite different from its physiological time.

The anthropologist Jordan (1993), reflecting on her comparative study of birth in Guatemala and the U.S., commented on how the third stage of labour was overlooked in her ethnographic note taking. Her familiarity with medical settings meant she did not notice the very different experience of this phase in rural Guatemala, which was only observed when viewing video records. In this setting, where blood loss or problems with placental separation could be life-threatening without access to medical treatments such as blood transfusion, it was clear that labour did not end until the placenta was delivered.

Furthermore, the clear marking of time and phases of labour which characterized hospital birth in London is not so present in a less medicalized setting, whether that be a home birth in a Western society or births in much of the non-industrialized world. We have already noted how the transfer from home to the hospital environment prompts a need to be able to define a 'proper', specific time for the beginning of labour, despite its gradual physiological nature. It also prompts a marked end, following the baby's birth and the rapid delivery of the placenta, at which point, if all is well, medical attention suddenly ends. Historically, midwives in most cultures followed women through the transition of birth and to the new family, with a strong emphasis on care following the birth, for up to forty days. In this setting, women were often shocked to find that all but the most basic care was suddenly withdrawn with transfer from the labour to the postnatal ward, with staffing and budget priorities being placed on the 'active' area of labour.[4]

### *Labour as Transition?*

The rapid transition the women described in modern maternity wards contrasts sharply with the historical tradition of 'churching' in the U.K. Following birth, women were secluded at home and expected to rest with their baby for forty days, with kin, neighbours or a 'monthly nurse' providing domestic care (Newell 2004). At the end of the forty days, the churching ritual, which marked the close of this period of liminality, effected a reincorporation into the ordinary social world with the new status of mother. Similar rites of passage surrounding the transition to parenthood are of major importance in many cultures (see, e.g., Kitzinger 1989; Vincent-Priya 1992). In the face of rapid social change and globalization – including the globalization of biomedicine – such practices have shown continuity as well as change. Hashimoto (Chapter 11, this volume) describes how the practice of *satogaeri* ('returning to old nest') for postnatal women has been adopted with shifts to the nuclear family structure in postwar Japan. Similarly, Donner (2005) described how demand for Caesarean birth among middle-class women in Calcutta has been influenced by changing norms and the desire among women to be able to return to their maternal home for rest and support following birth.

Referring to rites of passage, where the liminal phase is seen as being outside culture and society and thus fraught with danger, Douglas (1966) noted how ideas of risk were often applied to pregnancy and birth, a betwixt and between state. Rites of passage are not only found in traditional cultural rituals surrounding pregnancy, childbirth and the puerperium but are also found in Western hospital childbirth practices (Davis-Floyd 1994; Jordan 1993; Kitzinger 1989). In modern hospital birth, a woman is separated from her ordinary social world, with activities such as the removal of ordinary clothes, 'admission traces' and the withholding of food and drink, transferred to a labour ward for delivery and thence moved to a postnatal ward, where flowers and visitors are received. This parallels neatly the three phases of the traditional rite of passage, but the needs being served here are not necessarily those of the parents so much as the needs of the institution (and its desire to continue running smoothly), needs such as the staffing of wards and areas, keeping a throughput of beds by moving women quickly through the system, maintaining a sense of order through hierarchy and 'going with the flow' (Kirkham 1989; Hunt and Symonds 1995).

The time period of this medically oriented transition ritual is short, and increasingly so, the overwhelming interest of obstetrics being in active labour and the delivery of the baby. This is a transition period that bears a closer relationship to the concerns and interests of maternity professionals than to the significance of the transition in the woman's life, which is seen in a much longer time frame. The character of such rituals appears to be more to do with the needs of the institution and health professionals – who desire a sense of order and control in a situation characterized by uncertainty – rather than the needs of the woman and her family in this transition. Recent changes in the character and provision of postnatal hospital care bring out this difference particularly clearly, since the significance and meaning of the drama and journey of birth are barely acknowledged in the task-centred and limited character of postnatal hospital care (Simkin 1991; Ball 1994).

# Concluding Remarks

The midwife Jane Sharp, author of one of the earliest English texts written by a midwife, entitled a chapter of her book 'To know the fit time when the Child is ready to be born'. She warned:

> I shall desire all midwives to take heed how they give anything inwardly to hasten the Birth, unless they are sure the birth is at hand, many a child has been lost for the want of this knowledge and the mother put to more pain than she would have been. Let not therefore the child be forced out.... It is hard to know when the true time of her travel is near, because many women have great pains before the time of delivery comes. (Sharp 1999[1671]: 8)

She goes on to discuss the different signs of impending and actual labour, and warns against over eager interventions such as the breaking of the waters. Though written over three-hundred years ago, it highlights the fact that concerns with time are nothing new, nor are they just the province of 'medical men'.

Nonetheless, this discussion of women's experiences of time in labour has highlighted the tension between a reductionist, biomedical view of childbirth, and a broader bio-psychosocial view, as described by the anthropologist Jordan (1993) and echoed in these women's narratives. From the biomedical perspective, labour should be defined and thereby managed within a clear and limited time frame. This is motivated partly in order to manage risk and control the sense of uncertainty inherent in childbirth. It is also influenced by the imperative of the routines of a large institution, where, as Kirkham (1996) noted, midwives attend to the needs of the institution rather than to the needs of birth. That this view of risk management bears only limited relation to the underlying physiological theory of labour, or to scientific evidence on the effects of interventions, highlights that risk management is about broader concerns. The concern with risk management itself has developed within the context of the idea of 'risk society', as described by Giddens (2001) and others (e.g., Lane 1995). This suggests that the desire to mark and manage time in such particular ways has in itself contributed to the development of a risk-oriented culture. The dominant mode of hospital birth in late modernity has become one where risk management is increasingly employed to measure and hence control risk through attempting to reduce or eliminate the inherent uncertainty of the birth process. Controlling time features as a major aspect of the attempt to control uncertainty through mechanisms of measurement and control. However, the focus on risk management in itself heightens perception of risk and the need to manage it in this manner. As discussed in Chapter 2 (this volume), I would argue that this forms a dialectical process, each feeding into and reinforcing or constructing the other.

The nature of the risk to be managed is rarely a subject for discussion, despite the considerable policy and practice emphasis on risk in Western medical contexts. But it is evident that a wider, cultural approach to understanding ideas of risk, as with concepts of time, would be useful to further our understanding of the imperative of active management of childbirth. The irony is that the management of time in labour in medical settings does not appear to take into account growing evidence about the effects of environment and women's emotional responses to the physiology of childbirth, with the result that that conditions designed to speed up labour for apparent safety reasons may actually slow its progress and create other forms of risk which remain overlooked.

## Notes

1.  The active management of labour approach is set out in O'Driscoll, Meagher and Boylan (1993).
2.  The main risk is considered to be that of infection, but the midwives argued that risk of infection is actually increased in hospital when compared to home, and that signs of infection, which in practice is rare, can easily be watched for.
3.  Overviews of evidence have shown that the routine use of continuous electronic fetal monitoring may increase the risks of further medical interventions without any increased 'safety' benefit. However, the practice, once established, has proved difficult to change. In a study of the use of evidence around this practice, Beake (1999) found that practitioners referred to an unwritten set of guidelines which were more interventionist than the written guidelines that were present in the maternity unit.
4.  This lack of supportive postnatal care is discussed further in Chapter 10 (this volume).

## References

Albers, L.L. 1999. 'The Duration of Labor in Healthy Women', *Journal of Perinatology* 19(2): 114–19.

Allison, J. 1996. *Delivered at Home,* London: RCM Press.

Ball, J. 1994. *Reactions To Motherhood: The Role of Postnatal Care.* Cambridge: Cambridge University Press.

Beake S. 1999. 'What Factors Influence Midwives' Choice of How to Monitor the Fetal Heart in Uncomplicated Labour?' Masters dissertation. London: Thames Valley University.

Davis-Floyd, R. 1994. 'The Ritual of Hospital Birth in America', in J.P. Spradley and D.W. McCurdey (eds), *Conformity and Conflict: Readings in Cultural Anthropology.* New York: Harper-Collins, pp. 323–240.

Donald, I. 1969. *Practical Obstetric Problems,* 4th edn. London: Lloyd-Duke Medical Books.

Donner, H. 2004. 'Labour, Privatisation and Class: Middle-Class Women's Experience of Changing Hospital Births in Calcutta', in M. Unnithan-Kumar (ed.) *Reproductive Agency, Medicine and the State: Cultural Transformations in Childbearing.* Oxford: Berghahn, pp. 113–36.

Douglas, M. 1966. *Purity and Danger: An Analysis of Concepts of Pollution and Taboo.* London: Routledge and Kegan Paul.

Downe, S. 2004. 'Risk and Normality in the Maternity Services: Application and Consequences', in L. Frith and H. Draper (eds) *Ethics and Midwifery: Issues in Contemporary Practice*, 2nd edn. Oxford: Butterworth Heinemann, pp. 91–109.

Ehrich, K. 2003. 'Reconceptualizing "Inappropriateness": Researching Multiple Moral Positions in Demand for Primary Healthcare', *Health* 7(1): 109–26.

Giddens, A. 2001. *Sociology*. Cambridge: Polity Press.

Haddad, P. F., N.F. Morris and C.D. Spielberg. 1985. 'Anxiety in Pregnancy and its Relation to Use of Oxytocin and Analgesia in Labour', *Journal of Obstetrics and Gynaecology* (6): 77–81.

Hunt, S. and S. Symonds. 1995. *The Social Meanings of Midwifery*, London: MacMillan.

Huntingford, P. 1985. *Birth right: The Parents' Choice*, London: British Broadcasting Corporation, Continuing Education.

Jordan, B. 1993. *Birth in Four Cultures: A Crosscultural Investigation of Childbirth in Yucatan, Holland, Sweden and the United States*. Prospect Heights, IL: Waveland Press.

Kirkham, M. 1989. 'Midwives and Information Giving During Labour', in S. Robinson and A. Thomson (eds), *Midwives, Research and Childbirth*, Volume 1. London: Chapman and Hall, pp. 117–38.

——— 1996. 'Professionalization Past and Present: With Women or with the Powers That Be?' in D. Kroll (ed.) *Midwifery Care for the Future: Meeting the Challenge*. London: Ballière-Tindall, pp. 164–201.

Kitzinger, S. 1989. 'Childbirth and Society', in I. Chalmers, M. Enkin and M. Keirse (eds), *Effective Care in Pregnancy and Childbirth*. Oxford: Oxford University Press, pp. 99–119.

Lane, K. 1995, 'The Medical Model of the Body as a Site of Risk', in J. Gabe (ed.) *Medicine, Health and Risk: Sociological Approaches*, Oxford: Blackwell, pp. 53–71.

Leap, N. and B. Hunter. 1993. *The Midwife's Tale: An Oral History from Handy Women to Professional Midwife*. London: Scarlett Press.

Lewis, J. 1990. 'Mothers and Maternity Policies in the Twentieth Century', in J. Garcia et al. (eds), *The Politics of Maternity Care*. Oxford: Clarendon Press, pp. 15–29.

Llewellyn-Jones, D. 1982. *Fundamentals of Obstetrics and Gynaecology*, 3rd edn. London: Faber.

——— 1986. *Fundamentals of Obstetrics and Gynaecology*, 4th edn. London: Faber.

McCourt, C. 1998. 'Concepts of Community in Changing Healthcare: A Study of Change in Midwifery Practice', in I.R. Edgar and A. Russell (eds), *The Anthropology of Welfare*. London: Routledge, pp. 33–56.

——— 2006. 'Supporting Choice and Control? Communication and Interaction between Midwives and Women at the Antenatal Booking Visit', *Social Science and Medicine* 62(6): 1307–18.

Maubray, J. 1724. *The Female Physician*. London: James Holland, Bookseller and Printer.

Miller, A.W.F. and Hanretty, K.P. 1997. *Obstetrics Illustrated*, 5th edn. Edinburgh: Churchill Livingstone.

Newell, R. 2004. 'The Thanksgiving of Women after Childbirth: A Blessing in Disguise?' Ph.D. dissertation. Dundee: University of Dundee.

O'Driscoll, K., D. Meagher and P. Boylan. 1993. *Active Management of Labour*, 3rd edn. London: Mosby.

Sharp, J. 1999[1671]. *The Midwives' Book, Or the Whole Art of Midwifery*, ed. E. Hobby. Oxford: Oxford University Press.

Simkin, P. 1991. 'Just Another Day in a Woman's Life: Women's Long Term Perceptions of their First Birth Experience', *Birth* 18(1): 203–10.

Stallworthy, J. and G. Bourne (eds). 1979. *Recent Advances in Obstetrics and Gynaecology, No. 13.* Edinburgh: Churchill Livingstone.

Thompson, E.P. 1967. 'Time, Work Discipline and Industrial capitalism', *Past and Present*, 38: 56–97.

Van Gennep, A. 1960. *The Rites of Passage.* London: Routledge and Kegan Paul.

Van Hollen, C. 2003. *Birth on the Threshold: Childbirth and Modernity in South India.* Berkeley: University of California Press.

Van Teijlingen, E. et al. 2000. *Midwifery and Medicalization of Childbirth: Comparative Perspectives.* New York: Nova Science Publishers.

Vincent-Priya, J. 1991. *Birth without Doctors: Conversations with Traditional Midwives*, London: Earthscan.

Wilson, A. 1995. *The Making of Man-midwifery: Childbirth in England 1660–1770*, Cambridge, MA: Harvard University Press.

CHAPTER 10

# 'FEEDING ALL THE TIME': WOMEN'S TEMPORAL DILEMMAS AROUND BREASTFEEDING IN HOSPITAL

*Fiona Dykes*

## Introduction

This chapter focuses upon women's temporal experiences of breastfeeding while in hospital in England. It commences with an overview of the ways in which notions of linear time became a central aspect of the medicalization of infant feeding during the nineteenth and twentieth centuries. The predominant place of the clock in both culture and medicine was exemplified when the hospital became centre stage for infant feeding, and the forum within which rules and rituals related to linear time could be enacted. As the twentieth century progressed discourses around childcare and infant feeding changed to emphasize flexibility rather than rigidity, a shift that was exemplified in notions of demand feeding. However, changing medical discourses around recommended styles of infant feeding may clash with deeply embedded cultural norms and the realities for women engaging in such practices. This chapter draws upon an ethnographic study of breastfeeding women and midwives with regard to breastfeeding experiences while in hospital (Dykes 2004, 2006). It illustrates the clash of time frames for women between the irregularity and uncertainty of 'demand feeding' and the overarching cultural imperative to connect with clocks and linear time.

## Contrasting Time Frames

During the industrial revolution, in Europe, the notion of mechanical clock or linear time inexorably overrode cyclical, rhythmic time (Cipolla 1967; Kahn 1989; Adam 1992, 2004; Bellaby 1992; Helman 1992; Starkey 1992). Kahn (1989) compared the concepts of 'linear time' (also known as 'clock time'), 'historical time' or 'industrial time', and 'cyclical time', also referred to as 'organic' or 'agricultural time'. Clock time, Kahn argued, measured by the clock and pitched relentlessly towards the future, is centred upon the notion of efficient production and the factory, both of which are deeply embedded within Western capitalist societies. In contrast, Kahn suggested that cyclical time is a bodily, rhythmic time that is part of one's ontology and not separate and 'outside' oneself like linear time. It relates to the 'organic cycle of life' in which one is 'living within the cycle of one's own body'; it is a time that is 'cyclical like the seasons, or the gyre-like motion of the generations' (Kahn 1989: 21).

During the Victorian era in England, spanning most of the nineteenth century, the factory became central to the lives of a great number of people. It was an era marked by growing preoccupation with authority, discipline and obedience (Beekman 1977; Foucault 1981), as well as an obsession with the clock, routines and schedules related to the requirements of efficient factory production. The super-valuation of clock time reached into many aspects of the lives of English people to include the ways parents raised their children. As Millard argued, the clock developed an 'unparalleled position as a symbol of science, discipline, and the co-ordination of human effort' (1990: 217).

Scientific discourses around infant feeding at the turn of the twentieth century reflected the reductionist, mechanistic and dualistic assumptions of the Enlightenment and the growing medical preoccupation with supervising and regulating women's bodies (Palmer 1993; Carter 1995; Smale 1996; Blum 1999). Metaphors represented breast milk as a disembodied product that was produced in a mechanical way. The progressive urbanization of society and its associated effects of social isolation contributed towards women increasingly seeking and engaging with 'expert advice' in place of intergenerational and embodied knowledge. The growing number of 'expert' texts on infant feeding (e.g., Rotch 1890; Budin 1907; Vincent 1910; King 1913) reflected medical fears about chaos and a desire to impose controls, as illustrated in the following extract:

> In all cases the success or failure of maternal nursing must largely depend upon the way in which the practical details are carried out. Where the methods are haphazard, and the mother feeds her infant at all sorts of times, sometimes over feeding and at other times under feeding it, the results are always unsatisfactory, and the infant is constantly suffering from digestive disturbance in some form; while in other cases the effects are much more serious. Twelve hours after the birth the infant should be put to the breast and allowed to suck for two or three minutes. From this time to the time

that the breasts are freely supplying milk, the infant should be given the breast every four hours. (Vincent 1910: 40)

The mother is rendered largely invisible in this text with her breasts being discussed as if independent of her. She is, however, mentioned with regard to the requirement for her to follow practical details and the problems she would cause if she failed to do so. The reader is then directed to a table of feed frequencies, commencing with the first day and continuing for four to six months, with one feed per night permitted until the infant is twelve weeks old when it should be discontinued. This type of feeding regime continued to be advocated throughout the first half of the twentieth century. It was asserted that feeds should be meticulously and rigidly controlled in terms of time spent at each breast measured to the nearest minute (Fisher 1985). As Millard argued, the clock provided the key frame of reference, 'creating regimentation reminiscent of factory work, segmenting breastfeeding into a series of steps, and emphasising efficiency in time and motion' (1990: 211). The extent to which these new feeding rules were actually implemented by women will of course have been highly variable, depending upon social class, literacy and access to medical services.

The twentieth century brought with it a dramatic increase in hospitalization of women during childbirth and postnatal recovery. The hospital, like the factory, was a place in which the principles of linear time and associated routines became central. The scheduling of breastfeeding provides a classic example of the imposition of time constraints upon an inherently cyclical, rhythmical and relational process. As Simonds argued, the 'idiosyncratic rhythms of breastfeeding (determined by mothers and newborns) were obfuscated and mechanically regulated by an obsessively precise schedule' (2002: 566). Within several decades, however, these patterns of rigid timing were to be disrupted by a new set of discourses around 'demand feeding'.

## Demand Feeding

From the mid twentieth century, as Beekman (1977) noted, authoritative ideas related to parenting and children began to be challenged and changed. This shift related to changing views regarding the individual as the mainstay of democracy and a gradual emergence, in some circles, of an emphasis upon emotional interaction between mother and baby. The concept of demand feeding first appeared in the Western literature in the 1950s (Illingworth and Stone 1952) and it constitutes an example of changing attitudes. The notion of demand feeding represents a striking reversal of the authoritative knowledge referred to above, in which the scheduling of feeding was reified. In essence, it represents the removal of all restrictions upon breastfeeds in terms of frequency or duration, thus enabling the baby to feed whenever they want to.

Demand feeding is interchangeably referred to as baby-led or flexible feeding (Woolridge 1995; UNICEF 2001) but the term 'demand' is commonly used in practice. The concept came about with the recognition that if babies were given unlimited and untimed access to the mother's breast then they would be able to regulate their own calorific and nutritional requirements (Woolridge 1995). Demand feeding is also associated with an increase in the period of time over which women breastfeed (Illingworth and Stone 1952; Woolridge 1995). Emphasis upon demand feeding grew slowly until it became a firmly established concept in the 1980s, with a particularly strong emphasis developing over the last ten years as a key recommendation of the global 'Baby Friendly Hospital Initiative' (WHO 1989).

It is important to recognize that although discourses and practices around childcare and infant feeding changed during the second half of the twentieth century towards increasing flexibility they did not erase all discipline-based approaches. Indeed, as ideologies around 'natural' and baby-led approaches have grown they have been increasingly juxtaposed with counter-arguments and an upsurge of 'authoritative' texts arguing for a re-establishment of routines in child care and infant feeding. The childcare manual by Ford (1999), for example, advocates a return to the management of babies using hour-by-hour, day-by-day routines, representing a re-emphasis upon timing, precision and control.

The increasing sales of books such as Ford's (1999) suggests that there is a degree of unease among women in the U.K. and other Western communities about demand or flexible feeding. While the physiological basis for demand feeding is strong (see Smale 1996), there has been little reference to women's abilities and experiences of responding to their baby's needs in this way. As Smale points out, 'the vocabulary of the two main styles of breastfeeding, "demand" and "schedule" feeding, carry considerable emotive weight' (1996: 238).

While the distinction between scheduled and demand feeding appears to be clear there is considerable ambiguity in relation to the notion of demand feeding in texts which advocate it. For example, breastfeeding texts tend to refer to the need to allow a baby to feed without restriction but then go on to define the 'normal' range of the frequency and duration of feeds at specific stages following birth. A glance at the language in medical texts when referring to demand feeding illustrates the persistence of discourses that are distinctly time, transfer and measurement orientated. Take the following example:

> It is advisable for numerous reasons to feed young infants whenever they indicate a desire to feed. When left to their own devices, infants feed for greatly varying durations, with length probably determined by the rate and effectiveness of milk transfer. Infants who are permitted to regulate the frequency and duration of their feeds suckle more, gain weight more rapidly, and breastfeed for longer periods than infants who are restricted in their feeding patterns. (Saadeh and Akre 1996: 156)

The emphasis here is still upon the efficient transfer of milk from mother to baby, with the woman remaining invisible and any suggestion of mutuality and relationship absent. These time-orientated constructions of demand feeding are striking in their contrast to the ways of feeding seen in cultures in which babies are carried on their mother's abdomen with constant access to her breast. As Palmer (1993) commented, to ask a mother in some cultures about the frequency of breastfeeds would be like asking people how often they scratch when they have an itch.

Spiro's (1994, 2006) anthropological research with Gujarati women is illuminating here, especially in relation to cultural interpretations of time. She studied the meaning of breastfeeding for Gujarati women mainly living in Harrow, U.K. Women who had recently lived in rural communities in Gujarat had an agricultural, cyclical concept of time, related to the sun and the seasons rather than the clock. However, those who had lived longer in a Western culture developed a more linear concept of time, the extent of which was related to the length of time they had lived there. Spiro (1994) asserts that agricultural time, for Gujarati women, is rhythmical and seasonal with breastfeeding being part of the cycle of life. Childbirth and breastfeeding are seen as a time of rest or 'time out', with the woman's relationship with the baby being seen as a time of special intimacy, harmony and mutuality. This 'time out', experienced in rural communities around the world (Baumslag and Michels 1995; Vincent 1999), sharply contrasts with the experiences of many women in the U.K.

To advance her understanding of Gujarati women and time, Spiro (1994, 2006) drew upon Kahn's (1989) idea of a form of cyclical time called 'maialogical time'. Kahn developed the neologism maialogical time to refer to the period during a woman's life when she gives birth, raises children and breastfeeds them. She developed the word from the Greek *maia*, which means to mother or nurse (1989: 27). This word comes from the Indo-European root *ma*, which derives from the notion of the child's cry for the breast. She selected this word stem because it emphasizes embodied relationality and gives voice to both baby and mother. Kahn (1989) contrasted the concept of maialogical time with linear time, the former relating to the relational self and embracing interaction, mutuality, reciprocity and inter-relatedness. Linear time, she argued, is 'inhabited by individuated western man who follows the linear trajectory of history … its sociability is based upon the collective activity of "autonomous" individuals frequently in competition with one another, or working for the benefit of someone else at the expense of the self" (1989: 28).

Forman argued that the notion of maiological time would enable women to not only 'live *in* time' but to '*give* time' (1989: 7), a crucial influence on the way in which parenting may be perceived and experienced. Some would argue that women in Western industrialized cultures have become so programmed by linear time that they may be unable to embrace or experience cyclical time (Kahn 1989; Adam 1992). However, this view contributes to dualistic representations of living in and with time, suggesting that we can only engage with one form of time or another. Kahn (1989) disrupted this dualism by reflecting upon her own experiences of motherhood and feeding, which she argued allowed her to experience cyclical time in spite of living predominantly within linear time. In particular she referred to her own experience of returning to work, where linear time predominated, and then contrasted this with her experience of cyclical time, in the evenings, when she breastfed her baby.

Kahn (1989) suggested that women can experience aspects of maiological time through the experiences of pregnancy, birth and breastfeeding. However, she recognized that living in a culture where linear time dominates makes this a challenge for women. Balsamo et al. (1992) likewise argued that although breastfeeding may take a woman outside the industrial conception of time, trying to negotiate the two types of time can cause feelings of conflict and dissonance for women. The narrative data presented in this chapter, from women in the early twenty-first cenury, when scheduled breastfeeding is no longer formally advocated, reflects similar tensions.

## Critical Ethnography

The tensions for women with regard to demand feeding were explored in a recent ethnographic study in England (see Dykes 2004, 2006). The underpinning theoretical perspective stemmed from critical medical anthropology, with its focus upon blending macro and micro perspectives:

> [Critical medical anthropology] takes positions on the medicalisation of everyday life in contemporary society, which it opposes; on biomedicine as a form of power, domination, and social control, which it also opposes; and on mind-body dualism, again in opposition. ... [It] includes critique of medicine as an institution, cultural criticism focused on the domain of health, analysis of capitalism in the macro-politics of health care systems and the micro-politics of bodies and persons, addition of historical depth to cultural analysis, and critique of allegedly non-critical medical anthropology. (Csordas 1988: 417)

Critical ethnography constitutes a methodology that stems from the critical medical anthropology perspective. Ethnography involves participating in people's lives, including watching what happens, listening to what is said and asking questions (Hammersley and Atkinson 1995). It uncovers two types of cultural knowledge: tacit knowledge, a knowledge that remains largely outside our immediate awareness; and explicit knowledge, a form of knowledge that people may comment on with relative ease (Polanyi 1967; Spradley 1980). Critical ethnography places additional emphasis upon ideology, power and control in the research process, analysis and theoretical conceptualizations. As Thomas argued, it involves a:

type of reflection that examines culture, knowledge and action. It expands our horizons for choice and widens our experiential capacity to see, hear and feel. It deepens and sharpens ethical commitments by forcing us to develop and act upon value commitments in the context of political agendas. Critical ethnographers describe, analyze, and open to scrutiny otherwise hidden agendas, power centres, and assumptions that inhibit, repress, and constrain. (1993: 3)

The study referred to in this chapter explored the influences upon women's experiences of breastfeeding within postnatal ward settings. A critical ethnographic approach was adopted in two maternity units in the north of England, with sixty-one postnatal women and thirty-nine midwives participating. The ethnographic study involved long periods of observation of activities on the postnatal wards and interactions between midwives and breastfeeding women, and these observations were supplemented by interviews with midwives and breastfeeding women.[1] As Hammersley and Atkinson (1995) argued, participant observation and interviewing are mutually enhancing in that what was seen informed what was asked about and what was heard at interview informed what was looked for. Ethical approval for the study was gained through the relevant local research ethics committees, and at all stages participant autonomy and confidentiality were protected while pseudonyms are utilized in the quotes used below. This chapter refers to one specific aspect of women's experiences: temporality and breastfeeding.

## Tensions and Dilemmas in the Temporality of Breastfeeding

There is little emphasis in the literature on breastfeeding upon the ways in which women in a Western culture interpret, experience and negotiate demand feeding their babies and yet, in this study, demand feeding was crucial and central to women's experiences. Women were heavily influenced by their need to be in control, with linear time placing powerful limits upon their experiences and expectations with regard to breastfeeding. While in hospital they were coping with the past, the birth, the present with all its challenges and the future. The future was marked by the temporal notion of time moving on towards the re-establishment of 'normality' with a major part of that being related to returning to paid employment and an orderly life. Women's ways of negotiating breastfeeding in hospital therefore related to varying degrees of desire to be in control of their life both in their immediate situations and in the projected longer term.

The ambiguities regarding demand feeding created temporal confusion for women. They knew it involved a flexible approach – feeding the baby when the baby was hungry – but still often felt unsure about what this involved:

I don't know whether I'm feeding him enough. How long should you feed when they are demanding? (Helen)

I think in the night, it was more what I expected it to be, just sort of every four hours, but since eleven [this morning; it is now 5 p. m.] it's just been constant... I didn't expect that. It's all so contradictory, so many pros and cons; you never truly know. (Barbara)

### Unpredictability

The irregularity and uncertainty of demand feeding and the removal of culturally ingrained linear temporal markers from a lived and embodied experience created considerable discord for women, making them feel dislocated in time. Balsamo et al. (1992) referred to this as socialization into order within Western communities with unscheduled breastfeeding being seen as disorderly, time consuming and exhausting. The tensions women experienced in relation to the variable and unpredictable nature of demand feeding are illustrated by Lesley:

It's been a bit irregular and yesterday, I don't know whether it was because there were a lot of visitors around, but, he wasn't taking a lot. It was just little bits here and there and then I was winding him and putting him down and he wasn't settling and he was ... like ... he wanted more, so I was up till about 2 o'clock.

Lesley's remark about her baby taking 'little bits' reflects what Helman referred to as the Western linear assumption that 'every event or phenomenon will have both a beginning and an end' (1992: 37). In contrast, women felt confident if a baby stayed on the breast for a period of time which they felt was acceptable. For example:

I felt more confident once she'd actually latched on, and once she's there she tends to stay there. I think if she'd been mooching about and coming on and off all the time I think that would have made me really nervous. (Tracy)

Women tended to become anxious when there were changes in the 'pattern' of demand feeding:

OK, she's been feeding every few hours, but, she's not woken up since five so I'm feeling a bit like, well, not so confident. (Megan)

### Frequency and Duration

Most of the postnatal women and many of the midwives spoke in ways that indicated a strong orientation towards the clock, illustrating the deep embeddedness of linear time in Western lives. In spite of women referring to themselves as carrying out demand feeding they were preoccupied with the frequency and duration of feeds:

Well, she had a proper feed just before visiting hours ... and then another
feed just after visiting hours, like for about forty-five minutes, and she was
sucking really hard and that was on each breast. And then, about 12 o'clock,
from then on till half-five, she was ... feeding and then sleeping and waking
up and having more and that. Then they took her out; then she had about a
thirty minute one at quarter to seven, and after that she went to sleep until
about 10 o'clock. (Millie)

Millie's reference to a 'proper feed' relates to women primarily associating
breastfeeding with nutrition. It also illustrates the Western linear notion of
every event having a beginning and end (Helman 1992). Millie went on to
refer to bottle-feeding as the prefered norm in terms of the predictability
of frequency and duration of feeding:

I mean, like, the bottle-feeders, I hear them say, 'Oh, she's had this much
and they don't feed again for this many hours', but you can't really judge
with breastfeeding and that bothers me while she's so demanding.

While midwives advocated demand feeding, their language often reflected
the embeddedness of linear time within their ways of knowing: for example,
they often requested detailed information about the baby's frequency and
duration of suckling, asking: 'What time did he last feed?' 'And what
about previous to that?' 'How long did he feed for? This concern about
frequency and duration of feeds reflects Western associations between
time and quantity. As Adam stated: 'clock time, the organizational frame
and structure of industrial production, is governed by the non-temporal
principle of invariant repetition. Objectified and reified it is related to as a
quantity' (1992: 160).

Jenny, one of the midwives, highlighted some of the tensions for
midwives and breastfeeding women related to their personal attitudes to
demand feeding:

I think midwives tend to be guided by what worked for them. No matter
how many courses you go on, you tend to do what works for you. I think
demand feeding is one of the slowest items of all. Mothers come in with this
idea of four hourly feeds. They have that expectation and they're concerned
if the baby goes longer and they're concerned if they go more frequently.
I think that's part of the tension around breastfeeding. They have this
expectation of three or four hourly, timetabled feeding. And when a baby is
feeding on and off they think there's something wrong.

Jenny referred to the ways in which the philosophy of measurement of
feed frequencies and durations was at its most striking in paediatric advice
issued in case notes:

Demand feeding is an endless source of tension. For a normal baby, the
paediatricians have this idea of regular feeds, and they frequently write it
on the charts when called to delivery for something like meconium liquor
or low Apgar. Even when resuscitation hasn't been needed or has been
successful, they'll put: 'Plan: ward with mother, monitor temperature, four
hourly temps, three hourly feeds, early feeding'. It's like a mantra really, and

you've got a strong, healthy normal baby who doesn't need any particular regime at all. You know the paediatric chart with the tick list? When the paediatrician has been at a birth you'll see that on almost every one. It's the beginning of pathologizing.

This preoccupation with timed measurements to assess wellbeing, in addition to timing feeds, resonates with the suggestion by Thomas that 'time provides not only ways of describing the distribution of events but also a basis for interpretations and explanations' (1992: 65).

### Temporal Boundaries

Women appeared to perceive notions of demand feeding as problematic with regard to their expectations of a 'good baby' being one who limits their demands and sleeps for acceptable periods of time:

I've kept an open mind if it didn't work. Like, I know people who have breast fed, you know, for ages, with both their children. And I know another couple of people who just couldn't ... get the hang of it and just turned to bottles straight away. And their babies have been absolutely fine ... They slept and, you know, they were good babies. So I've kept an open mind so that I wouldn't be disappointed if he needs the bottle. (Sophie)

Tracy, meanwhile, saw good behaviour as not messing around:

Yesterday, I felt a bit [she uses a negative gesture] ... because she wasn't feeding, but then, as soon as she started feeding in the night, I felt OK. She either feeds or she doesn't, she's quite good. She won't just sort of mess with it all the time. (Tracy)

Women appeared to expect their babies to fit specific activities into bounded sections of time. This resonates with Helman's (1992) assertion that 'the clock – as a crucial organizing principle in industrial society – symbolizes control, conformity and co-operation in social and economic life' (1992: 43). This relates to the Western imperative to civilize the baby (Lupton 1996; Schmied 1998; Meyer and de Oliveira 2003). As Lupton remarked, 'mothers domesticate children, propelling them from the creature of pure instinct and uncontrolled wildness of infancy into the civility and self-regulation of adulthood' (1996: 39). Food and eating have constituted a key route to achieving this civility throughout recorded history (Fildes 1989; Maher 1992; Lupton 1996; Vincent 1999).

### Setting the Baby's Clock

Vincent emphasized that in the earlier decades of the twentieth century 'experts' advocated disciplining infants through an 'external schedule to accustom their nervous systems to certain types of food, rest and play' (1999: 53), and this was seen to prepare the infant for a scheduled, clock-constrained later life. However, Vincent argued that by the mid twentieth century the timetable was not simply seen as an external means of imposing discipline but as an innate characteristic of the child: 'The clock has moved from the realm of culture as perceived in science, training and discipline to that of nature and organic processes. It has moved from outside to inside the human body ... The clock having been internalised is now thought to be inherent in human behaviour' (1999: 54).

Vincent's assertions are illustrated in the data from the ethnographic study described in this chapter. Some women appeared to have the expectation that even if demand feeding was practised the baby should and would, after a short period of time, get into a routine. This may relate to their expectation that routines should be quickly learnt by infants. However, it seemed that some women had so internalized ideas about routine and regulation that innate programming to clock time was seen to be 'natural' and indeed expected.

Sophie expected a routine to develop, giving the impression that if this did not happen soon she would reconsider her feeding options:

> He's slept and settled, but it's my first day and sometimes they don't feed as much on the first day.... I'll see how I go on through tonight and tomorrow and see if he gets in a routine.... I think if he was in a routine I could feed him for twenty minutes [or] half and hour, and then – three hours later, four hours later – he'd take it again.

Kate appeared to see her baby moving positively in the direction of establishing a routine:

> She seems to be getting into a bit more of a routine and the last feed she didn't have as long on. She's had a bath this morning too, at eight-fifteen. Then she fed at nine and she's just fed now, for about twenty or thirty minutes.

Kate then set out to control breastfeeding through placing time constraints upon it and developing a routine:

> I mean, this demand feeding, it's OK to begin with but then I want to get her into some sort of a pattern by one or two months, you know, like six to eight weeks or something. Once she's settled into a routine I'll put her cot in the nursery we've got ready

Kate appeared to want to rapidly progress her baby towards independence with development of a routine being a key element of this goal. Smale observed a similar expectation among breastfeeding women she counselled, who desired 'that a period of total unpredictability would resolve into a set regime' (1996: 236).

While most women saw demand feeding as a transient phase that, in time, would resolve to conformity to external and indeed internal clock time, they did not always assume that a routine would 'naturally' develop. When there was doubt they tended to look ahead in time and worry about how demand feeding would 'fit in' with their lifestyle.

> What puts me off is the demand feeding. I've got horses and I like to be out with them. The idea of sitting and feeding all day isn't me. (Gemma)

## Babies Taking Time

The emphasis upon babies behaving and conforming to clock time conflicts with research around babies 'taking time'. The notion of babies taking time is referred to in both neonatal developmental studies – such as that of Meyer Palmer, Crawley and Blanco (1993), who outlined the stages in development of competent co-ordination involved in suckling and swallowing – and sociological literature – Kahn (1989), for example, argued that babies live in maialogical time, illustrating fundamental sociability from birth. She referred to the way in which a baby, after a non-disrupted birth, displays embodied interaction with the mother and makes their way to her breast and suckles. This facilitates placental separation so that the baby initiates their own realignment with their mother.

Hannah, who had several children, clearly understood the idea that babies take their time:

> I'd say for everybody to just bear with it, because a lot of people want to breastfeed. But because the baby won't take off them they tend to get upset as well. My friend was like that, but if she'd have just stuck with it, the baby would eventually.... She got very depressed with it because she'd really been looking forward to breastfeeding but her baby was constantly crying and was just taking a few sucks at the breast, but, if she'd stuck with it, it would have been all right.... She ended up bottle-feeding because she was thinking, 'Why doesn't the baby want to know?' I think that this issue should be put on leaflets and stuff, because its very frustrating if you want to do something but your baby won't do it. They need to point out that it's not just *your* baby that's like that, and that if your baby doesn't take to it straight away it's not that its discontented with you but it's just taking time. Just put the baby in [your] nightie and lie with them at night or when [you]'re cuddling them and they will eventually mooch.

Hannah's emphasis upon babies 'taking their time' was rarely referred to by breastfeeding women or midwives. Perhaps if women understood the concept of maialogical time they might feel less pressured about their babies 'taking their time'. However, the drive to return to 'normal' life and its routines as quickly as possible are in tension with the idea of maialogical time. Indeed, women appeared to feel that their lives were 'on hold' while breastfeeding. This sense of breastfeeding being short term, marginal, disruptive and indeed liminal is referred to by others (Schmied 1998; Mahon-Daly and Andrews 2002; Sachs 2005).

## Disembodied Progress

Despite the global recommendation that women exclusively breastfeed their babies for the first six months of life, and then combine appropriate complementary foods with breastfeeding up to two years of age or beyond (WHO 2003), the reality for women often stands in stark contrast. A combination of lack of confidence, fear of chaos, unpredictability and sense of planning for the future, led to women trying to maintain their boundaries and control breastfeeding through making plans to incorporate 'bottles'. This was commonly seen as a necessary and desirable progression towards independence for the child and a return to 'normality' for the woman with normality being a euphemism for independence for the mother. While notions of 'embodied progress' are referred to in relation to women's reproductive health and associated practices (see Franklin 1997), the concept of disembodied progress was more appropriate in this context.

When women conceptualized breastfeeding as simply a transferral or delivery of breast milk to the baby, this facilitated an easy step to the provision of breast milk without the baby being at the breast; that is, feeding the child breast milk without breastfeeding. Expressing breast milk and feeding by bottle enabled this shift:

> I used to express for Heather and freeze it into ice cubes, just for if you were going out. Or sometimes, if I was tired, somebody else could give her a feed by bottle. I think its good to get them used to a bottle, because once Heather decided she didn't want me anymore it was easier to put her on to bottles because she had had bottles, you know, whereas if they get established for too long on the breast it can be difficult to put them on a bottle. If you do want to go back to work or there's another reason for stopping breastfeeding, then you can have a problem. I plan to do the same for this baby. (Shirley)

The most common form of disembodied progress involved the plan to transfer the baby to bottles of infant formula milk. The feeding bottle has become a powerful symbol of babyhood in the U.K. and, almost without exception, women saw breastfeeding as a relatively short-term project, with three to six months being the longest stated duration. Returning to work was usually given as the reason for progressing to the bottle and the nature of women's work was often seen as incompatible with breastfeeding:

> Well I've got to go back to work in eighteen weeks, so probably I'll breastfeed until then. I work for an airline so it is just not practical to express milk. If they gave us longer maternity leave I would. (Harriet)

Mandy, meanwhile, planned to combine breast and formula feeding and then, probably at three months, 'progress' to predominantly or all formula feeding:

> I aim to give him three months start. I've read that in the evening you can give a bottle to help them sleep. It's thicker. I've spoken to others with practical experience – they have combined both. Then it may be a bit more difficult after three months, there'll be a few more restrictions, because I'm going back to work.

It was particularly striking that women spoke repeatedly about breastfeeding as if it were simply breast-milk feeding with little emphasis upon the notion of relationality with the baby.[2] This reflects the medicalized discourse in which the nutritional value of breast milk is super-valued and conceptually separated from the bond of sociality. It seems unsurprising then that the task was seen as demanding because the act of breastfeeding was conceptualized as a one-way transfer of nutrition. Without the two-way reciprocity of a relationship, the act of providing for another is likely to be experienced as depleting.

## Reconceptualizing Women's Time

The ethnographic material presented above illustrates that women's experiences of breastfeeding were heavily influenced by their relationship with linear time. Breastfeeding tended to be experienced as time consuming and as interfering with more pressing calls upon women's time. In a society in which productivity, in the industrial sense, is especially valued, and with lives structured around daily routines, breastfeeding women are faced with multiple contradictions. Approximately 70 per cent of U.K. women with babies currently return to work within nine months of birthing, with many returning much sooner. Therefore, combining paid employment, usually away from the home, with childcare is now a key issue for the majority of women.[3]

Women are now increasingly engaged in two forms of production, reproductive and paid employment, and this creates specific pressures upon them as they negotiate conflicting experiences of temporality (Balsamo at al. 1992; Blum 1999; Dykes 2002, 2005a, 2005b, 2006; Galtry 2003). Balsamo et al. summarized the tensions between the 'natural' time of women's bodily rhythms, flows and fluctuations and the social time of production:

> Milk comes as the contractions of labour come and then the child and before them all menstruation, breaking into patterns of social time. They have rhythms of their own, linked to the relationship of the women to her physical and social background. They constitute a disturbance to the organisation of labour. And thus 'natural' individual time and the 'social' time of production come into conflict through the body of the woman. This conflict is even more dramatic today because production times have been accelerated with respect to 'natural time', but also because women are becoming more and more integrated into the world of production and its forms of knowledge and are ever more dominated by it. (1992: 85)

If time is conceptualised as linear and related to efficiency and productivity, in the industrial sense, then women are likely to perceive and experience breastfeeding as time consuming and even time wasting. This resonates with Adam's reference to time as a finite and quantifiable resource with 'time running on and out' (1992): 162).

Fear of using up too much time was reflected in women's concerns about feed frequency and the length of time their babies slept. Women's anxieties about the disordered and time-consuming nature of breastfeeding, and the pressure upon them to get out into society and engage with production as paid employees, were evident within the first three days of their new, mothering experiences. Women therefore endeavoured to place controls upon the time taken in the act of breastfeeding. The conflict between women taking time out with their babies and the resumption of 'normal' life, including getting back to paid employment, was ever present. To address this tension women can now separate out their 'supply function' from the demands of their baby. They can express breast milk or give formula milk that may be fed to the baby by someone else, somewhere else. The baby's needs for nurture may be satisfied with a dummy rather than the breast. The marketing of the bottle, breast-milk substitutes and breastfeeding aids targets this 'modern' woman. Blum referred to the separation of the mother from breastfeeding in the U.S.A.

> The breastfeeding-wage-earning supermom ... is paradoxically free from any embodied constraints or wants. She is treated and treats herself as nearly body-less, and can be endlessly self-disciplining... Today's supermom gets medical approval to carry her breast pump to work, and, through her milk, to maintain her claim to exclusive, class-enhancing motherhood. (1999:183)

Kahn (1989) argued that linear time is now so deeply embedded within Western culture that other notions of time are difficult to conceptualize. She asserted that linear time is outside and separate from the body, whereas cyclical time is an embodied time that is a part of one's being. This differentiation between linear and cyclical time is useful as a conceptual lens for understanding women's experiences. However, the narrative data presented here suggests that linear time had indeed become a powerful part of women's own ontology. As Foucault stated, 'time penetrates the body and with it all the meticulous controls of power' (1977: 152).

Women's bodily rhythms and flows, and the maialogical time of their babies, contrasts with the socially dominant linear time of Western industrial societies. Kahn highlighted the tensions that this may create for women who are expected to parent their children within a social system dominated by linear time. She argues that linear time is 'extremely inhospitable to the slower tempo of children' and that there is little support or recognition for women who put in most of the 'time' in the care of children (Kahn 1989: 28).

The transformative project lies in seeking ways to reconceptualize time in relation to mothering and infant feeding. It seems that if some of the limits or constraints upon women's time were removed then their perceptions of breastfeeding might also change. Breastfeeding might no longer be seen as simply using time and taking time from other activities. Simonds argued for changing conceptualizations of time referring to the strictures imposed by the medical model's clock: 'Time is not only money, as the well-known aphorism claims. It is also power. If we take the time to reconsider these models, perhaps with time, demystification may lead us toward the reconceptualisation of procreative time and the enhancement of procreative experiences' (2002: 569).

Kahn's (1989) notion of maialogical time has possibilities for changing understandings of time and restoring time for women to breastfeed in more fulfilling ways. However, she warned against reviving essentialist notions in which women are designated to a full-time childcare role. She argues that:

> Uncorseting our maternal bodies does not have to be incompatible with living in linear time, providing that this time moves forward more slowly and with more digressions. Thus there would be time out for children.... Perhaps the time will come when both productive and reproductive labour will be honoured equally. Not the tokenism of Mother's Day, but an appreciation expressed through the reorganisation of work structures to accommodate the uncorseted maternal body (Kahn 1989: 31).

Political activity is needed to restore the possibilities for women to take 'time out' for parenting and breastfeeding, should they wish to do this. Extensions to U.K. maternity pay and leave, in line with EU guidelines, is a positive move that reduces penalties upon vulnerable low-paid, part-time workers, whose rights were often very limited. Prior to these changes a third of women returned to work before their eighteen weeks of statutory maternity leave ended because they couldn't afford to stay off work.[4] Improving maternity rights, pay and work-place flexibility through statutory processes has had a marked impact upon the duration for which women breastfeed within Scandinavian countries (Galtry 2003). Clearly such statutory recognition sends powerful messages related to valuing parenting.

However, providing women with 'more time' will not necessarily lead to an automatic reconceptualization of time. As Adam suggests:

> We need to lift time from the level of the taken-for-granted meaning to an understanding that knows the relation between the finite resource, birth-death and being-becoming, between chronology, the seasons and growth. We need to de-alienate time: reconnect clock time to its sources and recognise its created machine character. (1992: 163)

With this recognition, women, as Forman (1989) argued, could be encouraged to seek to subvert the power of time over their bodies by creating rhythms of their own.

## Conclusion

This chapter has illustrated the temporal dilemmas for women in a Western community when engaging in an embodied activity, breastfeeding. Demand feeding, as an activity, exemplifies the clash of linear and cyclical time for women and, indeed, healthcare staff. The removal of temporal markers through a lived, bodily experience often creates discord, confusion and uncertainty. Women's fears related to the inherent uncertainty embedded in the notion of demand feeding leads them to feel the need to place control upon the situation. Women's plans to return to a 'normal life' and paid employment influenced their feelings about breastfeeding. Even at the very early stages of parenting a new baby there was a linear sense of time running on and out.

Women experience discord between authoritative knowledge about ideal breastfeeding and inherent cultural ambiguities. Breastfeeding within a cultural milieu, and indeed macro-culture, in which pressures over linear time are magnified and possibilities for relationality limited, leads to women experiencing parenting as 'demanding'. While the predominant feeling about breastfeeding is that it is demanding it will continue to be seen as short-term, marginal and disruptive.

There is a strong global emphasis upon protecting, promoting and supporting breastfeeding (see WHO 1989). However, protecting and supporting breastfeeding requires a redefining of breastfeeding as a relational, socially valued activity, and this necessitates a reconceptualization of women's time. Maternity legislation needs to enable women to have the time and space to engage in breastfeeding as an embodied activity while still maintaining a career, should they plan to do this. Women would not then be obliged to abandon paid employment nor become disembodied career women with expressing milk or changing to infant formula being their only options. If women anticipated this 'time out' without financial and other forms of loss then their experience of breastfeeding might be very different. If women embraced the concept of maialogical time and were able to incorporate it, at least partially, into their lives, they might feel less pressured about their baby's need for time. Indeed, having time to work flexibly *and* experience relationality through breastfeeding should be a basic right of women in all societies.

## Notes

1.  The research involved participant observation of 97 encounters between midwives and breastfeeding women, while 106 interviews with breastfeeding women and 37 interviews with midwives were conducted. The study followed the guidelines set out in the Association of Social Anthropologists' research guidelines: see Association of Social Anthropologists of the U.K. and the Commonwealth. 1999. 'Ethical Guidelines for Good Research Practice'. Retrieved 4 March 2005 from: http://www.asa.anthropology.ac.uk/ethics2.html.

2.  A further important theme which is not explored here, but discussed elsewhere, is the notion of the breast as a sexual object, creating category confusion and conflicts for women. Dykes, F. 2007. 'Resisting the gaze': the subversive nature of breastfeeding', in M. Kirkham (ed.) *Exploring the Dirty Side of Women's Health*, London: Routledge.

3.  See: Maternity Alliance. 2003. 'All Change on Maternity Leave?' Retrieved 14 October 2003 from: http://www.ivillage.co.uk/print/0,9688,171090,00. html.
4.  Source: see note 2.

# References

Adam, B. 1992. 'Time and Health Implicated: A Conceptual Critique', in R. Frankenberg (ed.) *Time Health and Medicine*. London: Sage, pp. 153–64.
———— 2004. *Time*. Cambridge: Polity.
Balsamo, F., G. De Mari, V. Maher and R. Serini. 1992. 'Production and Pleasure: Research on Breastfeeding in Turin', in V. Maher (ed.) *The Anthropology of Breastfeeding: Natural Law or Social Construct?* Oxford: Berg Publishers, pp. 59–90.
Baumslag, N. and D.L. Michels. 1995. *Milk, Money and Madness: The Culture and Politics of Breastfeeding*. New York: Bergin and Garvey.
Beekman, D. 1977. *The Mechanical Baby: A Popular History of the Theory and Practice of Child Raising*. London: Dobson Books.
Bellaby, P. 1992. 'Broken Rhythms and Unmet Deadlines: Workers' and Managers' Time-Perspectives', in R. Frankenberg (ed.) *Time Health and Medicine*. London: Sage, pp. 108–22.
Blum, L.M. 1999. *At The Breast: Ideologies of Breastfeeding and Motherhood in the Contemporary United States*. Boston, MA: Beacon Press.
Budin, P. 1907. *The Nursling*. London: Caxton Publishing Co.
Carter, P. 1995. *Feminism, Breasts and Breastfeeding*. London: MacMillan.
Cipolla, C.M. 1967. *Clocks and Culture 1300–1700*. London: Collins.
Csordas, T. 1988. 'The Conceptual Status of Hegemony and Critique in Medical Anthropology', *Medical Anthropology Quarterly* 2(4): 416–21.
Dykes, F. 2002. 'Western Marketing and Medicine: Construction of an Insufficient Milk Syndrome', *Health Care for Women International* 23(5): 492–502.
———— 2004. 'Feeling the Pressure, Coping with Chaos: Breastfeeding at the End of the Medical Production Line', Ph.D. dissertation. Sheffield: University of Sheffield.
———— 2005a. '"Supply" and "Demand": Breastfeeding as Labour', *Social Science and Medicine* 60(10): 2283–93.
———— 2005b. 'A Critical Ethnographic Study of Encounters between Midwives and Breastfeeding Women on Postnatal Wards, *Midwifery* 21: 241–52.
———— 2006. *Breastfeeding in Hospital: Midwives, Mothers and the Production Line*. London: Routledge.
Fildes, V. 1989. *Breasts, Bottles and Babies: A History of Infant Feeding*. Edinburgh: Edinburgh University Press.
Fisher, C. 1985. 'How Did We Go Wrong with Breast Feeding?' *Midwifery* 1: 48–51.
Ford, G. 1999. *The Contented Little Baby Book*. London: Vermilion.
Forman, F.J. 1989. 'Feminizing Time: An Introduction', in F.J. Forman and C. Sowton (eds), *Taking Our Time: Feminist Perspectives on Temporality*. Oxford: Pergamon, pp. 1–10.
Foucault, M. 1977. *Discipline and Punish: The Birth of the Prison*. Harmondsworth: Penguin.
———— 1981. *The History of Sexuality, Volume 1*. London: Tavistock.

Franklin, S. 1997. *Embodied Progress: A Cultural Account of Assisted Conception*. London: Routledge.

Galtry, J. 2003. 'The Impact on Breastfeeding of Labour Market Policy and Practice in Ireland, Sweden and the USA', *Social Science and Medicine* 57: 167–77.

Hammersley, M. and P. Atkinson. 1995. *Ethnography: Principles in Practice*. 2nd edn. London: Routledge.

Helman, C. 1992. 'Heart Disease and the Cultural Construction of Time', in R. Frankenberg (ed.) *Time, Health and Medicine*. London: Sage, pp. 31–55.

——— 1994. *Culture, Health and Illness*, 3rd edn. Oxford: Butterworth-Heinemann.

Illingworth, R.S. and D.G.H. Stone. 1952. 'Self-demand Feeding in a Maternity Unit'. *Lancet* 5(1): 683–7.

Kahn, R.P. 1989. 'Women and Time in Childbirth and during Lactation', in F.J. Forman and C. Sowton (eds), *Taking Our Time: Feminist Perspectives on Temporality*, Oxford: Pergamon, pp. 20–36.

King, F.T. 1913. *Feeding and Care of the Baby*. London: Macmillan.

Lupton, D. 1996. *Food, the Body and the Self*. London: Sage.

Maher, V. (ed.) 1992. *The Anthropology of Breastfeeding: Natural Law or Social Construct?* Oxford: Berg.

Mahon-Daly, P. and G.J. Andrews. 2002. 'Liminality and Breastfeeding: Women Negotiating Space and Two Bodies', *Health and Place* 8: 61–76.

Meyer, D.E. and D.L. de Oliveira. 2003. 'Breastfeeding Policies and the Production of Motherhood: A Historical-Cultural Approach', *Nursing Inquiry* 10(1): 11–18.

Meyer Palmer, M., K. Crawley and I.A. Blanco, 1993. 'Neonatal Oral-Motor Assessment Scale: A Reliability Study', *Journal of Perinatology* 8(1): 30–35.

Millard, A. 1990. 'The Place of the Clock in Pediatric Advice: Rationales, Cultural Themes, and Impediments to Breastfeeding', *Social Science and Medicine* 31(2): 211–21.

Palmer, G. 1993. *The Politics of Breastfeeding*. London: Pandora.

Polanyi, M. 1967. *The Tacit Dimension*. London: Routledge and Kegan Paul.

Rotch, T.M. 1890. 'The Management of Human Breast Milk in Cases of Difficult Infantile Digestion', *American Pediatric Society* 2: 88–101.

Saadeh, R. and J. Akre. 1996. 'Ten Steps to Successful Breastfeeding: A Summary of the Rationale and Scientific Evidence', *Birth* 23(3): 154–60.

Sachs, M. 2005. 'Following The Line': An Ethnographic Study of the Influence of Routine Baby Weighing on Breastfeeding Women in the North West of England'. Ph.D. dissertation. Preston: University of Central Lancashire.

Schmied, V. 1998. 'Blurring the Boundaries: Breastfeeding as Discursive Construction and Embodied Experience'. Ph.D. dissertation. Sydney: University of Technology.

Simonds, W. 2002. 'Watching the Clock: Keeping Time during Pregnancy, Birth and Postpartum Experiences', *Social Science and Medicine* 55: 559–70.

Smale, M. 1996. 'Women's Breastfeeding: An Analysis of Women's Contacts with a National Childbirth Trust Breastfeeding Counsellor in England 1979–1989', Ph.D. dissertation. Bradford: University of Bradford.

Spiro, A. 1994. 'Breastfeeding Experiences of Gujarati Women Living in Harrow'. M.Sc. dissertation. Uxbridge: Brunel University.

———— 2006. 'Gujarati Women and Infant Feeding Decisions', in V. Hall Moran and F. Dykes (eds), *Maternal and Infant Nutrition and Nurture: Controversies and Challenges*. London: Quay Books, pp. 232–49.

Spradley, J.P. 1980. *Participant Observation*. New York: Holt, Rinehart and Winston.

Starkey, K. 1992. 'Time and the Hospital Consultant', in R. Frankenberg (ed.) *Time, Health and Medicine*. London: Sage, pp. 94–107.

Thomas, H. 1992 'Time and the Cervix', in R. Frankenberg (ed.) *Time, Health and Medicine*. London: Sage, pp. 56–67.

Thomas, J. 1993. 'Doing Critical Ethnography', *Qualitative Research Methods* 26: 1–35.

UNICEF. 2001. *Implementing the Baby Friendly Best Practice Standards*. London: UNICEF U.K.

Vincent, P. 1999. *Feeding our Babies: Exploring Traditions of Breastfeeding and Infant Nutrition*. Cheshire: Hochland and Hochland.

Vincent, R. 1910. *The Nutrition of the Infant*, London: Baillière, Tindall and Cox.

WHO. 1989. *Protecting, Promoting and Supporting Breastfeeding: The Special Role of Maternity Services*. Geneva: World Health Organisation and UNICEF.

———— 1990. *Innocenti Declaration*. Florence: World Health Organisation.

———— 2003. *Global Strategy for Infant and Young Child Feeding*. Geneva: World Health Organisation.

Woolridge, M. 1995. 'Baby-Controlled Feeding: Biocultural Implications', in P. Stuart-Macadam and K. Dettwyler (eds), *Breastfeeding: Biocultural Perspectives*. New York: Aldine De Gruyer, pp. 168–217.

CHAPTER 11

# LIVING WITH 'UNCERTAINTY': WOMEN'S EXPERIENCE OF BREASTFEEDING IN THE CURRENT JAPANESE SOCIAL CONTEXT

*Naoko Hashimoto*

This chapter was developed from my reflections on an ethnographic study about breastfeeding that attempted to understand the obstacles and positive aspects of breastfeeding from women's point of view. The study was designed in order to illuminate the gap between theory and practice regarding breastfeeding, the gap between the discourse 'breast is best' and the difficulties that women come across in their life settings. The study was conducted in Tokyo, where I was working as a community midwife. Six women were interviewed at home at approximately one-month intervals from their first postnatal visit until the baby's first birthday. The women's narratives represented the biophysical, social, cultural and personal aspects of child caring as a whole. As a result, breastfeeding was framed as part of a process of becoming 'a mother and her baby in tune'.

The theme of time became apparent in the early stages of my study whilst I was questioning the philosophical assumptions around 'ways of knowing'. The concept of research is based on Western philosophical traditions, and this made me think about the application of Western notions of scientific rigour – such as objectivity, generalization and the dualistic mind-body paradigm – across cultures. Coming from a non-Western background, I could not take these for granted. In addition, the field of my study, Japanese women, could not be understood in terms of Western cultural assumptions alone. In the Western anthropological tradition, cosmologies can be analysed for what they show about people's view of time and space. Culture, therefore, should be understood through the native's framework (Hendry 1999).

Comparing the Western, linear model of time and examining Japanese cosmology within its historical and religious context, the concept of 'multidimensional cycles' seems the most appropriate term to describe Japanese people's view of time and space, which is based on the following assumptions:

- Because nature is considered to be a limitless space, concepts of time and space are perceived as eternal and part of universal existence.
- Because human beings exist as part of nature, human life exists as part of the natural cycle.
- Because human beings and nature are interconnected, human life cannot be understood separately from the environment.

In my study, these philosophical assumptions were also examined in relation to midwifery: the personal view of time and space becomes the basic philosophy of practising midwifery, which defines the personal view of birth and death. Therefore it influences each midwife's way of working as a midwife. My ethnographic study was designed to draw on my philosophical position as a midwife, caring for women, their babies and families, and their environment as a whole. The Japanese expression *zen-jin-teki* (meaning 'whole' and 'human' in an adjectival form) could be used for describing my position as a researcher, understanding the phenomena as a whole and their relation to their own environment. I was concerned about the element of objectivity, which should be used to minimize my personal bias as a practitioner. I took the position of stepping back from the things that I believed I knew about breastfeeding. Soon I realized that the idea of stepping back is based on the Japanese concept of *kokoro* ('inner self'), which is different from the Western concept of objectivity, the idea of being detached from the world (Slife and Williams 1995).

The aim of this chapter is to discuss the complex nature of understanding breastfeeding in relation to time and space. I begin by describing the background to my study of breastfeeding, and then go on to discuss Japanese views of time and space. The historical and social context (including medical discourses) of breastfeeding, which creates the current social and cultural environment in which women experience childrearing, are then reviewed. The key theme of 'a mother and her baby in tune' is then discussed, and then the meaning of women's time and space in breastfeeding will be re-examined.

## Japanese Ways of Seeing, Knowing and Believing

### *Researching the Way of Researching Breastfeeding*

Reflecting back on my clinical experience of supporting women's breastfeeding, I perceived women's attitudes towards breastfeeding as shifting. Some women breastfed their babies in a very 'natural' manner, whilst others experienced difficulties. This shift was revealed in a Japanese national survey, which showed rates of around 45 per cent of women breastfeeding and 45 per cent mixed-feeding at one month after birth from the 1980s onwards, compared with rates of about 70 and 10 per cent respectively in 1960. This occurred despite national recommendations that 'exclusive breastfeeding is the best practice' (MCHWA 2004). It has been claimed that the current trend of mixed feeding is a modern social phenomenon, which appeared as a result of confusion and conflict between traditional Japanese and Western ideas about childrearing (Yamamoto 1983).

A gap between breastfeeding theory and practice has become an issue of promoting breastfeeding within a Western research context since the 1970s. Anthropological studies argued the importance of 'cultural learning' in breastfeeding, in which the knowledge and skills of breastfeeding were conveyed whilst women saw other women's breastfeeding in their own community (Vincent-Priya 1992). Hoddinott (1998) argued that breastfeeding should have been stored in women's actions, which is termed 'embodied knowledge' of breastfeeding. However, in cultures where the women came to know infant feeding by seeing other women's bottle-feeding, the women's embodied knowledge came to 'bottle-feeding'. In Western societies in which, by about the 1970s, bottle-feeding appeared as a cultural norm, breastfeeding was experienced as a problematic event (Scott and Mostyn 2003). In this context, the women were supported by theoretical knowledge rather than practical advice, which suggested that the society itself lost the embodied knowledge of breastfeeding. Those studies highlighted the importance of understanding women's experience of breastfeeding in relation to the nature of the field.

Maclean (1989) argued that the positivist paradigm has limited application for studying the complex nature of breastfeeding. Looking at the midwifery research area, breastfeeding was studied following either a biomedical model or a social science model alone, which also failed to study biophysical aspects and the impacts from the social and cultural environment together. In a Japanese context, Yamamoto (1983) argued that formalised knowledge was required once people lost their common-sense knowledge. He saw the increased number of lay-magazines and books about childbirth, parenting and feeding since the 1980s in Japan as a negative social phenomenon rather than a result of advanced scientific knowledge.

The difficulty of discussing the cultural aspects of breastfeeding was also identified. The term 'culture' is often used in a very ambiguous manner, whilst people use it to put a tag on everything, which hinders development of a profound discussion about the topic. When the same behaviour was found in different geographical areas, it was assumed by many writers to be imported from other cultures. Where the same behaviours appeared within the same location, it was termed a traditional culture (Kojima 1989). In the current globalisation, people came easily to access other cultural systems through media, and it became more difficult to clarify 'a local culture' (Kojima 1989). In addition, when the human behaviour came to be found as a people's 'habit', it could not be easily perceived as something special to look at (Bourdieu 1973; Edmondson et al. 2000).

Anthropologists argued that the language difference could be used for writers to illuminate the native's culture. To give a very simple example from my study, 'breastfeeding' is described as '*bo-nyuu*' in Japanese, with '*bo*' written in a Chinese character of 'a mother', and '*nyuu*' is 'milk'. Breastfeeding is about 'mother's milk' in Japanese, whilst breastfeeding in English focuses on the body part. The importance of translation work was further argued as 'a mode of thoughts', where the cultural differences should not be understood by word-to-word technical translation alone (Asad 1986).

As a result of the background work, my research became a study to research a way of researching breastfeeding. I framed my research through assumptions of a theory-practice gap: theory-practice gap would arise under conditions in which either knowledge itself was wrong or still some knowledge was missing (Sandelowski 1998, personal communication). However, I developed the view that theory-practice gap may arise when the knowledge is put into the wrong context or context was missing in the process of constructing knowledge. My ethnographic study was designed in order to identify 'context' in breastfeeding, in which the key findings emerged through the time and space that was shared with the women in the field. I set my role some extent as a historian: trying to make the sense of the current view of 'time and space' through past to present, and present to future.

### *Japanese Cosmology: The Position of Life, Nature, and Language*

In this study, the interviews were used as the main data resources. I perceived this as important to clarify Japanese people's way of seeing, knowing, and believing, as it features in Japanese communication. From the experience of living in two cultures, I became more aware of the philosophical differences influencing people's ways of expressing 'self'. Whilst the Western linear model of time seems to view human life as a single existence, the researcher's viewpoint of making sense of the data is focused on the concept of 'identity'. One's own life experience needs to be understood within one's own life-time, whilst it is based on the belief: the life has a clear beginning and a clear end.

Reflecting Japanese ideas of self, people are not likely to express 'self' in communications. Moreover people try to hide 'self'. Hendry (2003) argued that Japanese people's politeness existed in the level of ideology, which was strongly influenced by Confucianism. The cultural value system is complex, however, as it cannot be understood by any particular religious thought alone. I perceived that people's ways of knowing, seeing, and believing are also very much stored in people's everyday life, from which the features of 'multidimensional cycles' could be illuminated.

The elements of viewing time as a natural cycle could be explained by religious philosophies such as Shinto, Buddhism and Confucian thoughts. The origin of Japanese land and Japanese people was narrated in a Japanese archive written in *Wado 5* (Western year of 712)[1]. The story told of 'nature', which brought the eternal and universal energy to life. The first humans, called *Kami* (It translates as God, but in a Japanese sense, *Kami* means 'a human'), emerged from the earth (nature) as did language. The story told the equal position amongst human, environment, and language. Just as life emerged from the earth, death was described as the time when all lives go back to the earth.

The people's aim of life is considered as to take the role of living as a part of nature, which could be attained through living with nature in harmony. As people's emotions such as anger, jealousy, sadness, and over joy were considered to disturb the harmony, people try to respect others' and one's own environment, which appeared as Japanese people's 'politeness'. The relationship between human and environment is described as like the mirror and its reflection, like as a cause and effect. The idea of cause and effect is not about a simple one-to-one relationship. Whilst time and space is considered as an eternal and universal existence, people would never know when, how, and in which condition the cause will appear as an effect. Therefore Japanese people try not to show their own emotions to others, and 'silence' is used for showing their politeness to others.

The silence in Japanese communication is considered as the moment of 'reflecting self'. Japanese people considered every natural object such as a tree, a river, and a cloud moving in the sky, all have the language to talk to humans. The understanding of their messages is dependent on the people's ability of listening, which is based on the state of *kokoro*. In Zen Buddhism, the highest state of *kokoro* was described as 'selflessness' (Cohen 1994). Aida (1972) argued the space of *kokoro* made it possible for people to purify their own thought and to understand others through sensing others' feelings. People who talk a lot are treated with much suspicion due to the lack of *kokoro*. The sense of deep gratitude and grief was expressed by unspoken manner, whilst people respect each personal space and time. Therefore each person tries to reflect one's own self, and synthesise one's own feeling to others'.

Chia (2003) argued that in the Anglo-Saxon cultures, 'to know' meant describing 'why' rather than 'what' or 'how'. In contrast, in most Asian cultures, 'to know' is about doing or performing rather than describing 'what' or 'why' or 'how' (Chia 2003). In a Western context, the spoken part of language had power, whilst silence has more value in a Japanese context. The theoretical knowledge that could bring no practical sense is considered less valuable. Within the Japanese context, philosophy is stored in people's action, which is called 'wisdom' and people's reflections take the great part of it.

Considering the usage of spoken and unspoken language, Doi (1971), a Japanese psychiatrist, characterised Japanese society as having high dependency to others (*amae* in Japanese), in which the degree of unspoken language was dependent on people's relationship. When the relationship becomes closer, less language is used. Furo (2001) argued that Japanese communication took the form of 'disagreement avoidance', in which people tried to avoid inviting any conflicts during talk by using silence. The communication also took the pattern of 'turn taking'; whilst someone is talking, the other people do not interrupt. Hendry (1993) described Japanese culture as 'wrapping culture' and Japanese communication as a form of 'exchanging gifts': the people wrap their *kokoro* within the conversations and sent their emotions as 'a gift' to others. It was largely found in unspoken parts of communication but also in the three different modes of speech: *teinei-go* (formal language), *sonkei-go* (respect form), and *kenjyou-go* (modest form). The people have to choose the accurate mode of speech according to the context.

Japanese conversations are very complex, as people use multiple opinions and show their different faces according to the time and place, which is influenced by relationship with others. There are no covert or overt conversations. If somebody was around in the same space, people would be very aware of the presence of others and change the structure of conversations. This means that in research the interviewer is likely to appear as a part of the data, as the interviews would be constructed through the women's perception towards the interviewer and also the environment in which the interviews are undertaken.

*The Japanese People's Attitudes towards Time: Life as a Natural Cycle*

Hendry (2003) described the modern Japanese life as living in 'a ritual harmony', where certain rituals were practised within extended family or community through the year. Within Japanese cosmology, human life is described by the metaphor of the seasonal cycle, in which spring is the time to birth, through spring and summer people's life gained richness to understand joy and hardship of life, and winter is the time to go back to the earth. The Japanese rituals followed the idea of 'cycle' and to 'live in harmony with nature'. The condition of 'illness' is explained by Japanese concept of *ke-gare* (*ke* means 'natural energy or energy for life force', *gare* means 'to be weakened', the word implies a tree was going to die), which was the condition where people's energy fell into disharmony (Kitou 2002). The women's reproductive events such as menstruation, pregnancy and childbirth were all considered as a state of *ke-gare*, so women were given time and space to recover. The following rituals are carried out after birth to the baby's first birthday:

- *ubu-yu*: The ritual of baby's first bathing after birth.
- *oshichi-ya*: Seventh day after birth, to give a name for the new baby.
- *omiya-mairi*: One month after birth, the families with the new baby pray to the Shinto temple for the healthy growth of the baby.
- *kui-zome*: 100 days after birth, which is the ritual to prepare the food for the new baby, and prayed for the baby to grow up with the fortune of having food throughout life.

Viewing life as a cycle, the rites of passage such as marriage or childbirth or ageing or death could exist as a shared value within families. For example, childbirth is considered as the time when people learn about the meaning of parenthood. From experiencing its hardship, they came to know the gratitude towards their own parents. The family rituals are the time when young generations are able to learn the meaning of life. The Western individualism was introduced in the 1980s in Japan, and more and more people came to live in 'self-centred life'. However, whilst Japanese people live with those family rituals, young generations still have a chance to know the meaning of life based on the Japanese cosmology; 'living with others and nature in harmony'.

### Research Framework: The Researcher's Emotions and Japanese Empiricism

Before describing my research approach, I would like to address some historical issues that influenced the meaning of research within a Japanese context: especially narrative/ethnographic approach; how emotions are used in the process of knowledge creation.

Within the Japanese formal context, the Government of each era initiated study of the philosophies from China and India and Western Europe from 6 BC. In the modern era, European philosophies such as Heidegger's phenomenology and Heuristic approach, and social theories such as Marxist and Weberian theory were studied in the 1930s and 1950s. The positivist approach was adapted to use just as a method that accorded with the Japanese Militarism, legitimated by Japanese Shinto (Watanabe 1976). After the Second World War, the Japanese academics engaged with active discussion about 'Japanese identity', whilst the North American and European philosophies were thoroughly studied. Aida (1972) described the Japanese way of integrating multidisciplinary ideas into a way that sits with Japanese traditional values system as *'awase'* culture (*awase* means 'going together').

The high literacy culture is one of the features of Japanese culture, thus ordinary people are the front line of creating and using knowledge. The Japanese literature had the area titled 'everyday diaries', 'travel diaries', and 'short poems', which narrated the people's habit or local weather and the features of people's life. Those diaries were found in the family business such as farming and fishing, which recorded the weather and its relation to the cultivation of farming (Watanabe 1976). Narrative/ethnographic approach was established in the area of folklore in the early 1900s, whilst the formal scholarly activities were undertaken in order to develop a Westernised and modernised civilisation. At the same time, the people coming from the ordinary background actively published folklore monographs, which were based on a fieldwork approach, collecting narratives about local rituals, habits, and dialects. The analysis was characterised by understanding local people's life experience through reflecting the researcher's personal emotions. This narrative approach was widely supported so as to be read by ordinary people. This approach was called 'Japanese empiricism' in order to distinguish it from European empiricism due to the different usage of researcher's subjectivity in the process of knowledge creation (Watanabe 1976).

My approach to developing my research approach could be explained by the Japanese concept of *awase* culture, since it was developed from my careful examination of Japanese and Western ways of knowing. The emotions and the personal experience were considered as the core of knowledge creation. The Western way of knowing is used for enhancing the reasoning and establishing logic in the research process. The following elements were considered in order to conduct the ethnographic study of breastfeeding within a Japanese context:

- Women's experience of breastfeeding is to be understood in relation to each individual life context, and also within social and cultural context.
- The Japanese concept of *kokoro* is used for the space of reflection and synthesising each other's thoughts and feelings.
- Japanese communication is considered as a negotiation process. It suggests that the researcher could appear as a part of the data.
- The translation work is considered as the translation of 'a mode of thought'.
- In order to understand the unspoken and spoken language together, the data are analysed by a holistic reading and the narrative approach is used to represent the essence of wholeness of women's experience.

### The Social and Historical Background of Understanding the Current Breastfeeding Practice

In spite of Western influences, breastfeeding seems a cultural norm in the current Japanese context, as breastfeeding is not discussed as a matter of 'choice'. This section will discuss the discourse that created the recent social and cultural environment in which women experience their breastfeeding and child caring.

### *The Value of Breastfeeding*

Historically, Japanese society had the longest duration of breastfeeding amongst other Asian countries, because without the tradition of dairy farming, animal milk was not used for baby's nutrition (Macfarlane 1997). In European countries, breastfeeding was shifted into wet-nurses, animal milk, and powder milk. The form of shift as well as the length of time taken for the shift would influence the meaning of the current breastfeeding discourse (Macfarlane 1997).

In the Japanese case, breastfeeding by baby's own mother was the only way for the babies to survive. In rare cases, babies were breastfed by other women, which was called '*morai-jichi*' (*morai* means 'begging' and *jichi* means 'breast milk'); when the woman had not enough breast milk or in case of mother's death, the female relatives breastfed the newborn baby. Most societies kept the knowledge and the local rituals and wisdom to help the women produce enough breast milk (Sawada 1983). In Japanese cases, local food or breast massage or praying at the Shinto Temple or to natural objects such as trees or mountains has been widely practised.

In the traditional mutual system, childbirth took place at *sanya*, a small hut, which was owned by the community. The women and their babies lived there for the first few weeks after birth separately from other families, in which the woman could take enough sleep, no housework, and breastfeed their baby as they wanted. Female relatives brought food and helped her child caring as well as breastfeeding. Yamamoto (1983) argued the *sanya* system was different from the current hospital practice, as the women used their time and space in their own accord. Breastfeeding was used for the ritual of bonding the community members, which is called *chichi-tsuke*. A few weeks after childbirth, the baby was breastfed by several different relatives. They became a community-mother; thus this ritual gave the meaning of 'protection' for new parents and their baby.

Breastfeeding was found everywhere in Japan until the 1960s, whilst people's life was based on rice-farming. During the day, the baby was always carried on the mother's back. At night the baby slept with the mother, and the baby was breastfed as the mother tried not to disturb other family members' sleeping. This carrying and co-sleeping arrangement, *onbu* and *soine*, was useful to protect the baby from accidents such as falling, but also made it easy to breastfeed. *Onbu* was considered as the chance for the small baby to learn about the social manners, whilst they listened and watched other people from the mother's back.

### The Hospital Practice

Hospitalisation of childbirth in Japan was initiated by outsider's view of safer childbirth - GHQ (American General Head Quarters) - during Post War Reconstruction in the 1950s (Oobayashi 1985). Before the shift, the hospital was used for abnormal cases. As normal birth was moved into hospital, midwives who had worked as an independent practitioner in each community were also moved into hospital settings. Since the 1980s, 99% of women gave birth in hospital settings (Mothers' and Children's Health and Welfare Association 2004).

The hospital breastfeeding approach was developed following the North American model of hospital protocol, which includes babies-mother in separation, regular breastfeeding according to the clock, the baby's test weighing before and after breastfeeding, the usage of powder milk, and the mothers' donning a gown, washing hands, and cleaning their nipples by medical wipes (Fukuda 1996). The hospital feeding room became the space in which the hospital midwives taught the mothers about the skill of breastfeeding in a restricted manner (Pearse 1990). The hospital practice seems to be reflected in the rapid increase in the mixed-feeding rate, which rose to about 45% in the 1960s, and has shown no change since the 1980s (Mothers' and Children's Health and Welfare Association 2004). The following historical review gives the idea of possible barriers to changing the current rate of mixed-feeding.

Firstly, in 1959, a group of paediatricians advocated 'combined feeding'; 5–10% of infant total intake should be fed by additional supplements such as rice starch or glucose water (Takahashi 1996). This guideline was instructed to mothers through the baby's follow-up clinics in the public health service. Coming towards the 1960s, powder milk became mass produced as a result of technological advance, and the additional supplement was replaced by powder milk (Takahashi 1996). The 1972 WHO recommendation that babies should be exclusively breastfed until the age of three to four months was adopted by the Japanese Government. However, combined-feeding had been established as a norm by then, and it fitted with the women's idea of westernised lifestyle (Takahashi 1996).

Secondly, in the 1980s, Japanese medical research established a discourse of insufficient breast milk. The role of the baby's one-month follow-up clinic was described as for doctors to check for inadequately breastfed babies (Nanbu 1983). This was found to be 5–20% of the attendance, and doctors were to positively instruct mixed-feeding for those mothers (Suzuki 1983). Responding to the 1989 WHO/UNICEF Baby Friendly Initiative, a 'rooming in system' was recommended with the following protocols; twenty-four hours' observation of babies in a nursing room, and then babies were moved to each mother's bedside, the mothers being asked to wash hands, clean nipples, inducted on how to use bottles, and the baby's weight was tested before and after breastfeeding on the six days, with a right feeding method to be instructed accordingly (Hachiya 1998; Sueshige et al. 1998; Senno et al 1995; Takei 1998). Effectively, there was not much change from the practice used in the mother-baby in separation system. It was titled 'hospital breast management', which sounds unfriendly for the women and their babies.

Thirdly, the women who live in urban cities as a nuclear family are supported by the family aid *sato-gaeri* (*sato* means 'old nest' and *gaeri* means returning), in which the women return to their own parents' home for childbirth and stay there for at least one month. The traditional Japanese wisdom is practised in this setting such as prohibiting the women to touch water, as it makes women's body cold and hinders to have enough breast milk. The women are encouraged to eat and sleep properly, whilst their mother cooks local foods for having enough breast milk. However, this family mutual aid appeared to have a rather negative impact on breastfeeding in the recent context (Iwai and Kawayoshi 2001), since the women's mothers belonged to a period when mixed-feeding came to be a cultural norm. They knew the importance of breastfeeding theoretically, but encouraged their daughters to add powder-milk (Iwai and Kawayoshi 2001).

## The State of Health and Illness

Finally, I will give attention to the current Japanese medical system, which gives a clue to understanding the difficulties in making a change in the current breastfeeding practice. Lock (1987) argued that the modern Japanese medical system had unique features since medical doctors have worked as its stakeholders.

The Japanese national health insurance was launched in 1961. In a basic principle, the medical service should be non-profit. However, the

Japanese Medical Association had more power to structure the medical services rather than the Government (Lock 1987). This made the Japanese medical system work as an aggressive market, as the doctors decided the medical service, medical devices, and pharmaceutical companies that could supply drugs to each hospital (Lock 1987). As a result of the rapid economic growth in the 1980s, Japanese people were put into the position of 'health care consumers', who could demand high technology in medical services. At the same time, the traditional Japanese herbal practitioners or massage therapists or acupuncturists remained in people's high demand, which made the Japanese medical system pluralistic and competitive (Steslicke 1987). Hospital practice followed the idea of 'risk management'. Therefore powder milk and regular bottle-feeding was considered as the way that the hospital could avoid the unnecessary risk such as baby's weight loss, even though theoretical and practical knowledge demonstrated that it should not be risky for babies to be fed by only breastfeeding.

In a Japanese context, once new protocols are launched as a part of the health care system, they are not easily changed. I looked at mixed-feeding as a problematic phenomenon. However, comparing it to the U.K. situation, 45% of exclusive breastfeeding rate and 45% of mixed-feeding rate at one month could be read as positive phenomena: 90% of women practised some form of breastfeeding. The Japanese women's preference of breastfeeding, rather than bottle-feeding, seems related to the cultural aspects of breastfeeding, which became a main theme of understanding the embodied knowledge of breastfeeding in my ethnographic work.

### Time and Space in Breastfeeding: The findings from the Ethnographic Work

My ethnographic research started from exploratory work; interviewing three women about their experience of breastfeeding at three to four months in the postnatal period. I took the open interview approach, thus the opening question began with; how are you getting on after having a new baby? The women's experience informed me that breastfeeding still exists as a strong social expectation in the current Japanese context, which further characterised the women's experience as follows:

- Breastfeeding was narrated as a problematic event by the woman who shifted into bottle-feeding at the very early stage of postnatal period.
- Breastfeeding was not perceived as something special to talk about in the interview setting once it became a part of a woman's everyday life.
- Breastfeeding was not a subject to discuss whilst the woman was uncertain about her everyday child caring.

Consolidating women's narratives in chronological time, breastfeeding was experienced as women's shift in time and space, which includes the shift in family relationship, or leaving their job, and the experience of living in different places such as in hospital or spending some time at their parents' home. I identified the limitation of researching breastfeeding by one stage type of interviews as follows:

- The timing of three to four months would not be a good timing for the women to reflect on 'ongoing events'.
- My position as a researcher or a midwife would not be appropriate for the woman to talk about their personal experience.

Considering the nature of the topic and the limitation of the one stage type of interview, the women's experience of breastfeeding was followed-up by longer time span (at least a year) and at one-month intervals. With ethnographic fieldwork, I reflected my everyday experience as a community midwife, through which I could increase my cultural accommodation to develop a deeper understanding of women's experience.

### Breastfeeding; A Woman and Her Baby in Tune

In my follow-up study, I took an open-interview approach, which aimed for the women to be able to talk about their everyday child caring by their own accord. As a result, breastfeeding was not talked about a lot in the women's narratives once breastfeeding became a part of their everyday life. Breastfeeding was narrated as background information. Therefore the women's experience of breastfeeding needs to be understood through reading the women's interviews as a whole, including both from spoken and un-spoken parts of women's stories and from my observations of the women and their babies.

In my analytical approach, I set one of the women's narratives as a master case, whose experience was based on being a first time mother and 'exclusive breastfeeding' (Note: the term exclusive means only breastfeeding is practised besides introducing weaning food). In addition, she kept a diary to record their breastfeeding and sleeping so that I could use her diary as a resource to see 'shifts' in breastfeeding as well as other aspects of their everyday life. In the first interview at baby's one month, she described her feeling towards breastfeeding as follows:

> I started to breastfeed soon after birth. We (she and her husband) were just amazed and seeing her to suckle ... We did not think about breastfeeding ... but from the beginning we just feel ... we like it ...(Mrs M - baby's 1 month)

This is the only extract where she directly talked about breastfeeding in the period of following-up her experience till eighteen months of her baby's age. In the rest of the interviews, she did not talk about breastfeeding itself, its value or benefits. Breastfeeding came with the episodes of baby's allergy, her relationship with her friends or medical professionals, and also some events such as a family funeral.

In the final interview, I asked "what do you feel about breastfeeding?" as I wanted to see how she would reflect on her experience. She said:

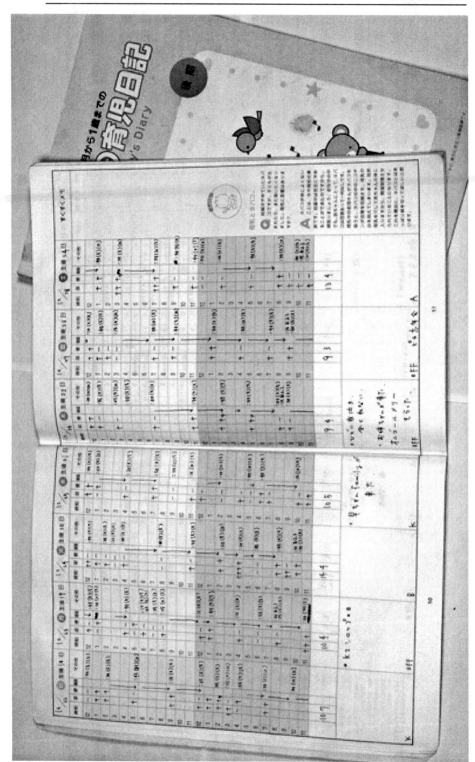

Figure 11.2: A baby-care and feeding diary.

I don't need to add anything ... I think you know about it ... Hospital staying? ... Yes, it was long enough to know about breastfeeding ... Breastfeeding is different from the things that I do once or twice a day ... Five days staying was enough to know about it ... (Mrs M - baby eighteen months)

Her comment clearly illuminated that her experience of child caring is about 'to know' about breastfeeding and 'to know' her baby. As her narratives told, she experienced a difficult time to adapt herself to a new life with her baby. However, she tried to do her best for each day, and her husband and her mother took the role of supporting her. Her transition seemed quite a smooth one, as I felt breastfeeding was embedded in her life from the first postnatal visit. However, following her narratives, I identified that the real sense of 'her and her baby in tune' appeared in the baby's fourth month.

Looking at her child caring diary, breastfeeding took place differently each day, as the woman described it as 'nothing she could feel it has patterns', which included the following features:

- Breastfeeding takes a form of interactive ways; not always tracing the form of one-way such as 'from many to less'.
- The interval of breastfeeding is different for each day.
- Night breastfeeding is also practised accordingly and also in a flexible manner; three to four times a night breastfeeding at times at the later stage is perceived as a normal part of baby's growth.

The women understood each day's breastfeeding followed the weather, her condition, and the baby's physical and emotional conditions. The women's narratives gave the sense of 'rhythm' in their life, the sense of 'control' and 'prediction'[2], and their baby-centred approach in breastfeeding. The following discussion aims to explore the meaning of 'a woman and her baby in tune' and possible elements that would make it possible for women to breastfeed in a baby-centred way, (rather than the scheduled feeding taught in the hospital) and its relation to Japanese cosmology; 'multidimensional time'.

### *The Women's Attitudes towards Breastfeeding; 'baby-centred feeding'*

In the process of translation work, I examined the concept of 'demand feeding' (see also chapter ten), which is generally described as *ji-ritsu-jyunyuu* in Japanese (*ji* means 'self', *ritsu* means 'regulating', and *jyunyuu* means feeding). The idea was introduced by Western child caring books, translated into Japanese in the 1970s (Yamamoto 1983). However, the traditional childcare told that breastfeeding should take a baby-centred approach, and the idea of demand feeding was not perceived as something new to the public, but led to confusion for the women (Yamamoto 1983). As most Japanese hospitals still used the system of baby-mother in separation, the women are likely to experience feeding by the clock in hospital settings. However, when the women are sent back home, the hospital staff tell them to breastfeed according to the baby's rhythms – *jiritsu-jyunyuu*.

In my study, the women's narratives illuminated the baby's crying at night, lack of sleep, and the difficulty to find the space for breastfeeding in public. During follow-up interviews, I saw the women's faces were very pale, which suggested their physical constraints from breastfeeding, especially for the first few months. However, the women never complained about breastfeeding, as they were persevering about child caring. I could not describe their breastfeeding as demand feeding as they did not feel they were 'being demanded by their babies'.

The women's early experience of breastfeeding was based on hospital task-oriented care, in which women need to follow the hospital protocol that prioritised the hospital business time. In fact, the women's experience of breastfeeding in hospital was much following the 'production-line labour' described in chapter one, in this book. The question must be; why or how women came to believe their sense, rather than to follow the legitimated knowledge of breastfeeding which they were trained to in the hospital setting. The women's narratives illuminated their transition from 'to learn' to 'to know', which is undertaken initially between the woman and her baby, but also the interaction between people around them. I identified two key themes to understand the process of transition:

- Women and their babies need time to develop their rhythms; they require about five months (at baby's fourth month) to come 'to know' each other's bodily rhythms.
- Women and their baby need 'space' in which they can immerse in breastfeeding; this time and space can be created whilst other people respect women and their babies' time and space.

### The Transition from 'to learn' to 'to know'

Medical anthropologists argued that breastfeeding was initiated by cultural learning, having been practised whilst the women were seeing other female relatives breastfeed. Medical anthropologists have also characterised human activities such as walking or eating as 'bodily experience', in which the human's bodily parts are engaged with the experience of the event. The action may be universal across cultures, but the meaning was found to be various as the manner of the action was constructed by the culture and social elements (Eriksen and Nielsen 2001). Breastfeeding could be considered as a bodily experience. Moreover, I clarified it as 'a bodily performance', where the action is taking place through interaction between women's and babies' bodily parts. The embodied knowledge of breastfeeding could be the knowledge learnt through bodily senses, which could be named 'bodily sensory knowledge' and cultural learning. Cultural learning could be described as about how societies or culture kept the way of learning bodily sensory knowledge.[3]

As I reviewed my own cultural way of knowing, Japanese culture seems to have the space to keep the way of learning 'bodily sensory knowledge', which is firstly found in the process for children to learn the manner of Japanese conversation: how to talk and how to not to talk. The space of *kokoro* encouraged people to 'reflect self' in silence. I would like to expand on the importance of silence in Japanese cultural learning, which often appears in learning Japanese traditional performances; *ka-dou* (Japanese flower arrangement), *sa-dou* (tea ceremony), *kyuu-dou* (archery), *ken-dou* (fencing). The term *dou* means 'knowing *kokoro*', the practise of which is based on the following three elements:

- The learner's capability of concentrating.
- The learner's sense of seeing and catching the key elements by seeing the teacher's performance.
- The learner's ability to know their own physical capability and limitation.

In this process, the learner is considered an active agent, whilst the outcome of learning is up to 'the learner's side'. The learner is asked to engage directly with the performance and repeat the performance until they could attain the stage of knowing the meaning of performance. The theoretical knowledge is used in the later stage for assisting the learners to reflect on their own performance.

In my study, I was concerned that all the women started breastfeeding with the mode of 'to learn' from hospital midwives. However, the women got into the stage of reflecting self by 'repeating the performance'. During the interviews, I was often with the women whilst they were breastfeeding. Once they latched the baby on their breast, I was waiting in silence. The women often said to me, 'it would not take long'. They seem to predict the length of breastfeeding from the baby's rhythm and strength of suckling. Spending some time on breastfeeding, they said, 'it will finish soon', feeling her baby's hands and feet getting warmer. I perceived that the women did not have this communication skill from the early stage of child caring. The process of knowledge creation started from the unconscious touching, and then the meaning was understood, which could be clarified as 'bodily communication'. I perceived that the women did not have bodily communication skill from the beginning. They also developed the sense of understanding baby's feeling through bodily contacts through breastfeeding. Once the women develop bodily communication with their babies, their life started to gain the sense of control and prediction; they could plan the time for going shopping or managing housework. This practice and sense of control was very different from that engendered by the routines of clock-timed breastfeeding taught in hospital and practised in bottle-feeding.

From my observations about breastfeeding and mixed-feeding women's breastfeeding performance at home, their action shows the different level of bodily reflections. Breastfeeding women performed breastfeeding as 'one completing action', which made them immersed in the moment of bodily reflection. In contrast, mixed-feeding women have several issues they need to be concerned with during breastfeeding; looking at the clock, checking their breasts, and thinking about how much formula milk the baby will need after breastfeeding. Mixed-feeding women did not seem totally immersed in breastfeeding, which would hinder how they could listen to their own bodily reflection as well as their babies. I was also concerned that their anxiety hinders their bodies being totally relaxed, which further influences the hormone release that relates to the 'let-down reflex' .

### Uncertainty and the Meaning of Control in a Japanese Context

I found the problem of hospital fragmented task-oriented care is to hinder the women's time and space to reflect self. In the final interview, one mixed-feeding woman said:

> ... No, No. Hospital stay was not long enough to learn breastfeeding ...
> (Mrs H - Baby's thirteenth month)

Reflecting back on her experience, her experience of hospital breastfeeding practice was described as follows:

> ... I enjoyed talking to other mothers in the feeding-room. I can see some of the women have massive breast milk from the beginning ... I checked her (baby's) weight before and after (breastfeeding), and then the midwife came to me and expressed my breasts. I can see only one or two drops of breast milk from me ... and the midwife said, 'nothing'. The bottle was topped up with formula milk, and then I fed it to her ... on the day of leaving, it shows 20g (after breastfeeding), and I felt I may be able to breastfeed, but the midwife said like ... 'it is nothing' ... I just feel ... I need mixed-feeding, because I have not enough breast milk ... It is *'shikata-ga-nai'* ... (Mrs H - baby's age two months).

The midwife's word 'nothing' put her in a passive position in reflecting self. As a result, when I first met her, she stopped thinking about breastfeeding as her own matter. The condition of not having enough breast milk was accepted as *shikata-ga-nai*, (*shi* means 'following', *kata* means 'rules', *nai* means 'not be able to something') which happened as a part of natural law.

In Japanese cosmology, the concept of *shikata-ga-nai* is based on viewing a human body as a part of nature. Thus people perceived uncertainty as a part of natural law, which they need to live with. The concept gives the negative connotation in Japanese conversations, which shows the people's helplessness when they need to give up their own expectation. However, it gives a rationale amongst people, which creates the space to accept the nature of limited capability of the human body. Thus people can make a shift into the next step.

Corbin (2003) argued that the concept of bodily limitation was not considered within Western research contexts, whilst the researcher's main focus was on the capability of the human body, as the Western social expectation of human-beings is to live everyday life with the sense of control. In everyday life situations, people should know their bodily differences and bodily limitations. However, the Western notion of power and control somehow is not taken into account in the research context. On the other hand, Jitsukawa (1997) argued that the sense of control in the Japanese context was found in the women's preference to stay close to nature or natural rhythms. In her study of women's experience of menopause, Japanese women did not like to use Western medical treatment such as taking pills as it interferes with the natural cycle (Jitsukawa 1997). In my study, the mixed-feeding women's narratives also gave the sense of control in their life, by continuing breastfeeding, which indicated that the women had the disposition to live close to nature or natural rhythm. The Japanese women's preference for mixed-feeding, not directly shifting to bottle-feeding, as is common in Western countries, was perceived as a cultural phenomenon.

In fact, within the U.K. context, national surveys show that mixed-feeding is not considered as a good option, since it required more time and energy. The women tended to stop breastfeeding swiftly, once mixed-feeding was introduced. In addition, most women have ceased breastfeeding by baby's six weeks of age (Hamlyn et al. 2000). It could be presumed that their experience could not reach to the stage of knowing the sense of control in breastfeeding. As a result, the women's experience of breastfeeding appeared as a problematic event, which could discourage other women from choosing breastfeeding.

### The Meaning of Support: A Japanese Concept of 'mi-mamoru'

The women's narratives illuminated that the meaning of support in breastfeeding is about protecting women's time and space in which women and their babies could immerse in breastfeeding. The idea was described as *mi-mamaoru* (*mi* means 'seeing' and *mamoru* means 'protecting') in Japanese, which is based on the mutual understanding between people. From talking to one of the women's mothers, the concept was described as follows:

> ... I know my daughter does not like to ask the things to me. I know how my daughter thinks ... She wants to try by herself first... I just need to wait and to see what she finds. I just need *mi-mamoru* ... I was very impressed her husband supports her really well. So now I think ... this is the way of young couples to do the things ... (Mrs K's mother - baby's eighth month)

The sense of *mi-mamoru* includes that the other people respected the individual bodily difference and limitation as the women perceived them, and helped the women to know their own body. I further identified that the concept is based on Japanese cosmology; viewing human existence as a part of nature, so that the meaning of support is also about respecting the uncertain nature of natural law.

The modern Japanese society and culture seemed heavily influenced by Western ideas, and the idea of a linear model of time started to appear as a part of Japanese value systems from the 1980s (Tsunoyama 1984). This shift may threaten the Japanese philosophical position and mutual understanding of others such as *mi-mamoru*. However, the women in my study showed their ability to reflect self, which further enabled them to develop the embodied knowledge of breastfeeding. The final discussion will be undertaken to reflect back on the role of cosmology in understanding people's experience including biophysical experience such as pregnancy or childbirth.

### Discussion: The Concept of Embodiment and Breastfeeding as a Craft

My ethnographic study informed the rhythm in breastfeeding from women's point of view, which illuminated the importance of understanding its biophysical nature, social and cultural elements, and also bodily performance, bodily reflection, bodily communication, and individual bodily limitation as a whole. In this section, I will look at two issues; the gap in the period of time that the women required to make a transition into motherhood, and the researcher and practitioner's philosophical position to understand the wholeness of human experience.

The closure of my ethnographic work was made by undertaking a final interview, asking each woman about the experience of breastfeeding. As a result, the women said, 'breastfeeding is good', which informed me that breastfeeding was initiated by women's feelings. Moreover, most of them did not put any particular value on their efforts or achievement of breastfeeding. The women's main interest was ongoing everyday life. I understood that the women's attitudes of 'seeing now and future' were commonly found in a Japanese context, which is based on 'multidimensional time'. I did not make sense of their experience in relation to the Western concept of identity, as it did not fit with the women's views of life.

I also identified the importance of reading/feeling/understanding 'silence' in women's time and space through the *zen-jin-teki* approach. Kenmochi (1978) argued the meaning of silence as *ma* in Japanese, which could deliver the meaning of time and space in one concept. The Japanese concept of *ma* could be translated into eight English words; space, interval, pause, room, time, while, leisure, luck, and timing. The Chinese concept of *'yin and yang'* was still based on the dualistic paradigm, whilst the Japanese concept of *ma* gave the third space for reflecting everything in its wholeness.

The women's narratives further illuminated the meaning of being 'with women'. In a research context, a longitudinal approach has been criticised as collecting snap shots, where the researcher could not fill the gap between the intervals. Using the concept of *ma*, I could bring my position to understand the women's message reading between the intervals. In my study, I was not always with them, but my monthly visit gave the sense of my presence in their life, as they were keeping their ideas and questions for my next visit, and phoned me and asked for some advice. The women also described the experience of living with a new baby as 'walking in the never ending story', which reflects their feelings of living each day with uncertainty. In this situation, my visits were perceived to give them rhythms to live with their uncertainty.

In order to convey the *zen-jin-teki* approach to the Western context, the sociological concept of embodiment was considered. In my study, the concept of embodiment could take two roles. Firstly, it is used for understanding women's experience as a whole. The women's narratives illuminated breastfeeding as biophysical conditions, emotions and feelings, their local ritual and personal belief, and then the embodied knowledge of breastfeeding was found in actions; the women and their babies' performing breastfeeding. In this context, women and their babies are both considered as active performers, often referred to as having agency. In medical discourses, the women's bodies tended to be discussed as in a passive position, described as a reproductive machine (Martin 1987).

Secondly, recent Western sociological discussions have illuminated the limitation of understanding the uncertain nature of human experience by a dualistic paradigm, which was hugely based on a disembodied approach (Williams and Bendelow 1998). Thus the idea of embodiment made a big shift in thinking, in which the human body has the subject voice to express its existence, so that research requires an approach that is able to listen to the voice. Csordas (1999) argued the concept of embodiment as able to define the researchers' philosophical position of being-in-the-field, as it could be the means of understanding other people's experience with the sense of attachment with others and also with their own environment. Therefore the concept of embodiment could change the direction from a dualistic approach; filling the gap between body/mind, subjectivity/objectivity, reasons/emotions, nature/culture, and male/female in separation. Within breastfeeding research, the concept of embodiment gives the possibility to integrate concepts as a whole that otherwise appeared in a dichotomy between nature/culture, emotions/reasons, science/art, and theoretical knowledge/intuition.

In order to illuminate the physical, emotional, and social and cultural aspects of breastfeeding together, I introduced the third concept of craft, which was used to discuss the bodily nature of medical practice (Carmel 2003). The concept of craft could fill the gap between culture/nature and science/art dichotomy of breastfeeding, which characterises breastfeeding as follows (Hashimoto 2006):

- Breastfeeding is practised in a flexible manner; no fixed pattern is found in each performance.
- Breastfeeding is performed using the common skills and knowledge, but the problem is uniquely experienced.
- The woman's bodily reflection gained through the interaction with her baby is used as the most convincing resource by which the woman comes to know about her own performance.

The concept of craft, along with 'baby's fourth month' and 'respecting women's time and space' could be used to develop the practical implication:

If the women could see the same midwife, every month, at least until the baby's fourth month, talking about breastfeeding, and knowing their breast conditions, they are more likely to develop their bodily senses to know their body and their babies. (Hashimoto 2006)

I found the concept of craft and supporting breastfeeding for a longer time period is challenging for the cultural assumptions and the current health service. Firstly, I considered the current Western theoretical knowledge dismissed discussion of the bodily nature of human experience. In the hospital setting, the discussion was taken on teaching the theoretical knowledge, benefits of breast milk, and the actual practice is initiated based on a temporally controlled manner; just teaching the correct techniques and skills of breastfeeding. The bodily aspects of breastfeeding, time required by the women to know breastfeeding and the amount of physical work required by women and their babies to come to 'know' about breastfeeding, were dismissed.

Thirdly, the baby's fourth month is also challenging, as the current maternity services are rarely able to provide a continuous carer who could work from a community base and also for longer than baby's four months. In addition, the Western notion of 'back to normal' gave a limited view to understand the transition. McCourt (2006) argued that the Western maternity service was initiated with the strong expectation for a swift transition to motherhood so postnatal visits were made for ten days by community midwives and until forty days by health visitors. The prolonged transition was considered as signifying a difficult mother, and the women's anxiety was diagnosed as postnatal depression. As a result, women were afraid to express their true emotions to medical professionals (McCourt 2006). In my research, the women's transition did not take such a swift manner; women required at least five months for developing the sense of being in tune with her baby. In this period, the women's emotional responses should be considered a normal part of women's transition. Therefore people need to listen to any woman's emotional responses as a part of transition.

The concepts of embodiment and craft could enable a shift in talking about the bodily nature of knowledge in breastfeeding but also in other areas of midwifery practice. My ethnographic work and narrative representation contribute to making a shift in researchers' philosophical position; each individual experience could enrich the picture of breastfeeding, and identify the elements that were dismissed in the previous knowledge. Especially Western breastfeeding research came to the stage of identifying the limitation of researching breastfeeding by a dualistic approach, and also the difficulties in clarifying the embodied knowledge of breastfeeding within a Western cultural context, where bottle-feeding exists as women's embodied knowledge. Therefore my study demonstrated the significance of cosmology in understanding human experience, which suggests the embodied nature of human knowledge could not be understood through analysing the social system alone or generalising from many cases by variables.

In my ethnographic study, I experienced living in two cultures. Last (2001), a medical anthropologist, described his dilemma in the other cultural field as follows:

... Nowadays anthropologists try to wear the native's framework as wearing contact lenses, realising its itchiness, irritated, and feel uncomfortable to wear them ...

This illuminated my discomfort; trying to understand the Western people's philosophical assumptions such as linear model of time and objectivity and subjectivity per se in Cartesian thought, but still I could not take it for granted. The language difference was experienced as a part of my discomfort: the more I came to know the profound difference, the more I came to believe I would not become a native English speaker. As a midwife, I perceived everyday practice is based on understanding women and their babies through women's lenses, therefore during my study I considered myself flexible to work between the different cultural ways of knowing, instead of being inhibited by my feeling of discomfort.

As I experienced, it is not easy to understand other cultural value systems. However, my study could suggest the possibility of a narrative/ ethnographic approach in which researcher's reflections, such as feeling, emotions, theoretical knowledge, clinical experience, personal belief, and time and space sharing with the people and spending in the field, could be used to convey the theoretical knowledge into practical understanding of the phenomena. I perceived the increase in social problems in Japanese child caring. However, the women who I came across in this study showed their passion and perseverence, wishing their babies to grow healthier, and offering their personal experience to share with other mothers. I hope that the women's narratives could be used for researchers and practitioners to reflect on their own practice as well as their way of understanding others. It will further create the space to exchange the cross-cultural dialogues about the different ways of knowing, and to reconsider personal philosophy; approaching each individual experience as a rich resource to learn the knowledge and wisdom of human life.

## Notes

1. As in China, a new emperor's reign in Japan signalled a new era of time, reflecting the degree to which the counting of time is a political matter.
2. I explore below how this sense of control and prediction is not necessarily the same as one would describe for a 'Western' woman. These are not rigid concepts, as the following discussion will illustrate.
3. This is well expressed in Bourdieu's concept of habitus.

## References

Aida, Y. 1972. *Nihonjin-no-Ishiki-Kouzou (Japanese: Japanese Way of Knowing)*. Tokyo: Koudanshya.

Asad, T. 1986. 'The concept of cultural translation in British social anthropology', in J. Clifford and G. Marcus (eds), *Writing Culture: The Poetics and Politics of Ethnography*, California: University of California Press, pp. 141–64.

Bourdieu, P. 1987. *Outline of a Theory of Practice*. Cambridge: Cambridge University Press.

Carmel, S. 2003. *'High Technology Medicine in Practice: The Organisation of Work in Intensive Care'*. Ph.D. dissertation. London: London School of Hygiene and Tropical Medicine.

Chia, R. 2003. 'From Knowledge-Creation to the Perfecting of Action: Tao, Basho, and Pure Experience as the Ultimate Ground of Knowing', *Human Relations* 56(8): 953–81.

Cohen, A.P. 1994. *Questions of Identity: An Alternative Anthropology of Self Consciousness*, London: Routledge.

Corbin, M. J. 2003. 'The Body in Health and Illness', *Qualitative Health Research* 13(2): 256–67.

Csordas, J.T. 1999. 'Embodiment and Cultural Phenomenology', in G. Weiss and H.F. Haber (eds). *Perspectives on Embodiment: The Intersections of Nature and Culture*. London: Routledge, pp. 143–62.

Doi, T. 1971. *The Anatomy of Dependence*. London: Koudanshya.

Edmondson, R. and C. Kelleher. 2000. *Health Promotion: New Discipline or Multi-Discipline?* Dublin: Irish Academy Press.

Eriksen, H.T. and S.F. Nielsen. 2001. *A History of Anthropology*. London: Pluto Press.

Fukuda, M. 1996. 'Rooming-in, Breast Milk and Breastfeeding (Japanese)', *Shyuusanki-igaku* 26(4):521–24

Furo, H. 2001. *Turn-Taking in English and Japanese: Projectability in Grammar, Intonation, and Semantics*, New York: Routledge.

Hachiya, S. 1998. 'Promoting Rooming-in: The Case of Kitami Red Cross Hospital (Japanese)'. *Jyosanpu-Zattushi* 52(10): 59–63.

Hamlyn, B. et al. 2000. *Infant Feeding 2000*. London: Stationery Office.

Hashimoto, N. 2006. 'Women's Experience of Breastfeeding in the Current Japanese Social Context: Learning from Women and their Babies'. Ph.D. dissertation. London: Thames Valley University.

Hendry, J. 1993. *Wrapping Culture: Politics, Presentation, and Power in Japan and Other Societies*. Oxford: Clarendon Press.

Hendry, J. 1999. *An Introduction to Social Anthropology*. London: Sage.

Hendry, J. 2003. *Understanding Japanese Society*, 3rd ed. London: Routledge Curzon.

Hoddinott, P. C. 1998. 'Why Don't Some Women Want to Breast Feed and How Might We Change Their Attitudes?' MPhil dissertation. Cardiff: University of Wales.

Iwai, Y. and K. Kawayoshi. 2001. 'Grandmothers' Attitudes to Breastfeeding and Their Influence on their Daughter's Breastfeeding (Japanese)', *Jyosanpu-Zattushi* 55(6): 72–78.

Jitsukawa, M. 1997. 'In Accordance with Nature: What Japanese Women Mean by Being in Control', *Anthropology and Medicine* 4(2): 177–99.

Kearney, M.II. and J. O'Sullivan. 2003. 'Identity Shifts as Turning Points in Health Behaviour Change', *Western Journal of Nursing Research* 25(2): 134–52.

Kenmochi, T. 1978. *MA-no-Nihon-Bunka (Japanese: the meaning of MA in Japanese culture)*, Tokyo: Koudan-Shya.

Kitou, H. 2002. *Bunmei-toshiteno-Edo (Japanese: Analysing the Edo era as a cultural system)*.Tokyo: Koudanshya.

Kojima, H. 1989. *Kosodate-no-Dentou-wo-Tazunte (Japanese: Reviewing the history of Japanese child caring)*, Tokyo: Shinyousya.

Last, M. 2001. 'Medical anthropology in the 21st Century', panel discussion *5th International Medical Anthropology Conference*, London: Brunel University.

Leap, N. 2000. 'The Less We Do, the More We Give', in M. Kirkham (ed.), *The Midwife-Mother Relationship*, London: MacMillan Press, pp. 1–8.

Lock, M. 1987. 'Health and Medical Care as Cultural and Social Phenomena', in E. Norbeck and M. Lock, (eds). *Health, Illness, and Medical Care in Japan*, Honolulu: University of Hawaii Press, pp. 1–23.

Macfarlane, A. 1997. *The Savage Wars of Peace*. Oxford: Blackwell.

Maclean, H. 1989. 'Implication of a Health Promotion Framework for Research on Breastfeeding'. *Health Education* 3(4): 355–60.

Martin, E. 1987. *The Woman in the Body: a cultural analysis of reproduction*, Boston: Beacon Press.

McCourt, C. 2006. 'Becoming a Parent', in L. Page and R. McCandlish (eds), *The New Midwifery*, 2$^{nd}$ edition. Oxford: Elsevier, pp. 49–71.

Mothers' and Children's Health and Welfare Association. 2004. *Maternal and Child Health Statistics of Japan'*, Tokyo: Mothers' and Children's Health Organisation.

Murphy, E. 2000. 'Risk, Responsibility, and Rhetoric in Infant Feeding', *Journal of Contemporary Ethnography* 29(3): 291–325.

Nanbu, H. 1983. 'Promoting breastfeeding (Japanese)', in H. Kato (ed). *Bonyuu-Hoiku*, Tokyo: Medica Science, pp. 158–69.

Oobayashi, M. 1985. *'Japanese Midwifery after the Second World War (Japanese)'*, Tokyo: Keisou-shyobou.

Pearse, J. 1990. 'Breastfeeding Practices in Japan', *Midwives Chronicle and Nursing Notes*, October: 310–15.

Raphael, D. 1976. *The Tender Gift: Mothering the Mother; the Way to Successful Breastfeeding*, New York: Schocken Books.

Sawada, K. 1983. 'An Anthropological View of Breastfeeding (Japanese)', in H. Kato (ed.), *Bonyuu-Hoiku*, Tokyo: Medica science, pp. 34–45.

Scott, J. and T. Mostyn. 2003. 'Women's Experience of Breastfeeding in a Bottle-Feeding Culture', *Journal of Human Lactation* 19(3): 270–77.

Senoo, T. et al. 1995. 'The Management of Breast Care and Protocol for Child Caring in the Baby Friendly Hospital (Japanese)', *Perinatal Care* 14(11): 1025–32.

Slife, D.B. and R. Williams. 1995. *What's Behind the Research? Discovering Hidden Assumptions in the Behavioural Sciences*. London: Sage.

Steslicke, W. E. 1987. 'The Japanese State of Health: a Political-Economic Perspective', in E. Norbeck and M. Lock (eds), *Health, Illness and Medical Care in Japan*, Honolulu: University of Hawaii Press, pp. 24–65.

Sueshige, K., A Nakao and Y. Ito. 1998. 'Promoting Rooming-in, in a Hospital Context of Breastfeeding Practice (Japanese)', *Jyousanpu-Zattishi* 52(10): 55–58.

Suzuki, S. 1983. 'Breastfeeding and Medical advice (Japanese)', in H. Kato (ed.), *Bonyuu-Hoiku*, Tokyo: Medica Science, pp. 501–508.

Takahashi, E. 1996. 'The Social Background of Breastfeeding (Japanese)', *Shyuusanki Igaku* 26(4): 459–64.

Takei, T. 1998. 'Breastfeeding and Breast Care in the Soka City Hospital (Japanese)', *Jyosanpu-Zattushi* 52(10): 38–43.

Tsunoyama, S. 1984. *Tokei no Shyakaishi* (Japanese: The Concept of Time in a Japanese Social Context). Tokyo: Chyuukou shinsyo.

Vincent-Priya, J. 1992. *Birth Traditions and Modern Pregnancy Care.* Shaftesbury: Element Books.

Watanabe, M. 1976. *Nihonjin-to-Kindaikagaku (Japanese: The History of Modern Science).* Tokyo: Iwanami Shinsyo.

Williams, J.S. and G. Bendelow. 1998. *The Lived Body: Sociological Themes, Embodied Issues.* London: Routledge.

Yamamoto, K. 1983. '*Bo-nyuu (Japanese: Breastfeeding)*', Tokyo: Iwanami-Shinsyo.

# CONCLUSION

The chapters in this book have looked at the theme of time and childbirth in various ways and from different perspectives. However, they all show ways in which methodological and theoretical approaches from anthropology can contribute to applied research on, in this case, maternity care and midwifery. The case studies presented in the chapters bring to life the very real concerns and dilemmas which link theory to practice. Here, we are literally touching on matters of life and death. Each has also touched to some degree on key debates within anthropology about the balance between universalism and relativism in social theory, and about the status of ethnographic approaches to research with respect to authoritative knowledge. A variety of perspectives, including feminism and critical theory, have been used as tools to think with, and in the development of analytical and critical approaches to what may be very familiar practices to the professionals who generally write about them. The majority of the contributors to this volume are practitioner-researchers. Consequently, they have taken on the challenges of subjectivity and objectivity as well as other familiar concerns within anthropology, about theory, reflexivity and validity in ethnographic research, and the issue of 'making the familiar strange' that lies at the heart of doing 'anthropology at home'.

A thread running through all the contributions is the fundamental standpoint they have taken of viewing all systems of health and healthcare as culturally situated and shaped, whether they be 'traditional', 'folk' systems or the universalizing system of biomedicine. The chapters have explored in various ways how systems of healthcare profoundly influenced by, if not centred on, biomedicine operate in practice with respect to childbirth and reproduction. They illustrate well the argument that reproduction is always a cultural as well as biological matter, and one that involves a continual interplay between physiological and psychological processes, and the natural and social environment, each of which can be described as complex and interactive systems.

The writers have challenged tendencies to view systems in terms of dichotomies; for example, treating the subjects of research in terms of oppositions such as male:female, science:nature, modernity:tradition. Nonetheless, taking a critical approach often involves making use of such oppositions. Several authors argue for complexity theory as a way of bridging and moving beyond such dualisms, while maintaining a critically engaged position. They argue convincingly that research in this area must engage with nature and culture, and it must be able to both consider relationships within systems and analyse specific aspects of practice, including clinical practice, as being complex and embedded within their context. In some chapters we have seen examples of time being treated in highly conventionalized, commodified or ritualized ways in maternity care, as well as examples of time as an embodied, complex physiological process, what the anthropologist Jordan (1993) might describe as bio-psycho-social. We have also seen particular case studies of struggles about ideologies and practices, struggles over who controls time, indeed who defines it, and how they do so, in childbirth and reproduction.

The anthropologist Bloch (1989) criticized more culturally relativist anthropological theories as confusing ideology with cognition, and some anthropologists for emphasizing the ritual and symbolic aspects of culture over the mundane. He argued that hierarchical cultures give particular emphasis to ritual representations of time, with ritual forms that tend to obscure the realities of the structures of power that lie behind them. In such contexts, ideological forms of time, such as a ritual calendar, come to dominate practical and embodied forms of time. We have shown that Bloch's argument is applicable to healthcare practices in resource-rich and complex societies as much as it is to the smaller-scale social systems that are often the focus of anthropology.

It is perhaps useful to say that rather than particular conceptions of time coming to dominate in hierarchical societies, a feature of such societies is the impetus and apparatus to shape these through power, authoritative knowledge and ritual. This book has presented case studies of how hierarchical social systems have been expressed and enacted in biomedicine and in the management of childbirth, and of how fundamental models of time and the management of time have been in this process. Biomedicine was developed primarily in Europe but its influence has been far wider, its universalizing tendencies paralleling those of colonialism and, later, globalization. We have given cases of how linear models of time are dominant in biomedicine, as applied to childbirth, and of reform projects which attempted to re-establish complexity or more cyclical perspectives. A few years ago, the anthropologist Mary Douglas argued that 'any institution that is going to keep its shape needs to gain legitimacy by distinctive grounding in nature and in reason: then it affords to its members a set of analogies with which to explore the world and with which to justify the naturalness and reasonableness of the instituted rules' (1987: 112). As we have seen, both biomedicine and its critics may appeal to reason and to nature to claim the ground of authoritative knowledge and practice.

The debate in anthropological theory about the importance of 'nature' or 'culture' is not new and is well known, and it has extended widely into other disciplines. Take, for example, theory of art, where writers and philosophers of art have discussed the question of 'physis' versus 'thesis', nature versus convention. Gombrich (1966), for example, emphasized the importance of culture and convention in the ways in which pictorial images are constructed by the artist and seen by the viewer. As cultural relativism became more established in the postmodern era, he restated his position to argue that visual images use both nature and convention, and that the codes by which people interpret images are not purely arbitrary but grounded in nature as well as shaped by culture (Gombrich 1999). This forms an apt comparison with the themes we have explored here about childbirth and time. We have suggested that much anthropological theory is concerned with the complex interplay of nature and culture, universalism and relativism. Looking back historically, one can see points at which anthropologists (often reacting to the social and political mores of their time) argued for a highly relativist position, while others have argued for a basic universalism that makes cross-cultural understanding, and the discipline of anthropology itself, possible (Bloch 1989). In this view, culture and cultural differences are important but not absolute, but they elaborate upon, shape and act upon nature.

In this book's chapters we have seen examples of ways in which medical procedures and practices, and medical knowledge itself, are highly influenced by convention, and in which what constitutes 'science' or 'evidence' may be a fertile ground for epistemological argument. A number of practices have been examined in depth, for which there is – even in terms of what is considered to be authoritative scientific evidence in evidence-based medicine – little evidence of measurable benefits and considerable evidence of iatrogenic risk, yet those practices continue because of their established nature, ideologies, relationships and structures of power. Ritual is often taken, in popular and medical thought, to mean something that is empty or lacking in (specific) power (like a placebo), and associated with atavism and religious and 'pre-scientific' thinking. In research on healthcare, 'ritualistic practices' are often cast as a methodological problem of the contamination of evidence, an aspect of context or behaviour that gets in the way of proper scientific enquiry. Some recent research, however, is beginning to acknowledge the need to understand the embedded nature of medicine in its ideas and practices – or what Bourdieu (1977) would refer to as dispositions – to discuss the limits of some forms of explanation, and to integrate this understanding into research knowledge and practices (Lambert and McKevitt 2002; McCourt 2005; Wrede et al. 2006). Just as Winter and Duff, in Chapter 3 (this volume), talked about the value of the midwives they studied being able to immerse themselves in, and be absorbed by, the women's embodied process of labour, so ethnography provides a method by which researchers can be immersed in and absorbed by the context of enquiry, while at the same time being able to step back to achieve forms of explanation which are not easily available by other means.

Our shared argument, therefore, is first that understanding childbirth needs other forms of research, which can view authoritative or scientific knowledge and practices as complex, contingent and socially situated, but also rooted in observations of and interactions with (universal) nature. Second, that research on childbirth, and research on concepts of time, can bring together understandings of the natural, the social and the cultural, in complex and fascinating ways that can help us advance (and be informed by) wider theoretical questions about the 'science' and 'art' of healthcare, and human nature and culture. Time in childbirth is physiological, psychological and social in the way that it is experienced and framed. The ways in which time in childbirth is conceptualized and managed enact social and cultural relationships of power and, in turn, help to confirm and institutionalize them. Such relationships have been shown to be enduring and deeply embedded, and obstetric approaches (such as active management of labour), and the power relationships between different maternity workers and their clients, have been spread universally as authoritative practice, putatively underpinned by scientific knowledge, by reason and nature. However, they are not static. They have been and will continue to be challenged. We have shown through various cases that the marking and management of time in childbirth has always been a concern, historically and in different contexts, and has always been culturally framed. We have also shown how obstetrics takes a particular, highly controlling approach to time in childbirth; it treats time in childbirth as universal even though it is socially situated. This controlling approach expresses power and it may generate risks as much as it manages them. We hope that this book will prove a useful contribution to debates about how time in childbirth should be managed in the future.

# References

Bloch, M. 1989. 'The Past and the Present in the Present' in *Ritual, History and Power: Selected Papers in Anthropology*. London: Athlone Press, pp. 1–18.

Bourdieu, P. 1977. *Outline of a Theory of Practice*. Cambridge: Cambridge University Press.

Douglas, M. 1987. *How Institutions Think*. London: Routledge and Kegan Paul.

Gombrich, E.H. 1966. *The Story of Art*. London: Phaidon.

——— 1999. 'The History of Anniversaries: Time, Number and Sign', in K. Lippincott et al. (eds), *The Story of Time*. London: Merrell Holberton, pp. 240–45.

Jordan, B. 1993. *Birth in Four Cultures: A Crosscultural Investigation of Childbirth in Yucatan, Holland, Sweden and the United States*. Prospect Heights, IL: Waveland Press.

Lambert, H. and C. McKevitt. 2002. 'Anthropology in Health Research: From Qualitative Methods to Multidisciplinarity', *British Medical Journal* 325: 210–13.

McCourt, C. 2005. 'Research and Theory for Nursing and Midwifery: Rethinking the Nature of Evidence', *Worldviews on Evidence-Based Nursing* 2(2):1–9.

Wrede, S. et al. 2006. 'Decentred Comparative Research: Context Sensitive Analysis of Maternal Health Care', *Social Science and Medicine* 63: 2986–97.

# NOTES ON CONTRIBUTORS

**Gisela Becker** completed her principal midwifery education in Berlin, Germany in 1986. She has since practiced in a variety of settings in Germany, the Caribbean and Canada where she has practiced as a midwife in Nunavut, Alberta, and Québec; and since 2000 the Northwest Territories. Gisela completed a M.A. in Midwifery Practice at Thames Valley University in 2006. Her dissertation topic was Women's Experiences of Culturally Safe Birthing with a Midwife in a Remote Northern Community.

**Soo Downe** spent 15 years working in various clinical, research, and project development roles at Derby City General Hospital. From January 2001 Soo has worked at the University of Central Lancashire in England, where she is now the Professor of Midwifery Studies. She set up the UCLan Midwifery Studies Research Unit in October 2002. Her main research focus at present is the nature of, and culture around, normal birth.

**Margie Duff** is a midwife with over 30 years experience in clinical midwifery, management and midwifery education, including working in Australia and New Zealand. In Australia she worked in rural and metropolitan hospitals. Since moving to New Zealand she has been a lecturer and programme manager for the Bachelor of Midwifery programme at Wintec. Margie completed her Ph.D. at the University of Technology in Sydney in 2005. She currently undertakes online education and research activities in Australia for the Charles Darwin University and the University of Western Sydney.

**Mavis Kirkham** is emeritus Professor of Midwifery at Sheffield Hallam University. She has carried out midwifery research and clinical midwifery since qualifying as a midwife in 1971. She sees herself as hybrid researcher, having worked in the disciplines of social anthropology, sociology, politics and history, using qualitative, observational and ethnographic methods as well as surveys and archive work. She has an abiding research interest in peoples' relationship with, and efforts to control their work and its setting. Her research on informed choice in maternity care led on to analysis of the culture of maternity services and the impact of the structure of services on service users and providers.

**Fiona Dykes** is Professor and Director of the Maternal and Infant Nutrition and Nurture Unit (MAINN) at the University of Central Lancashire. Fiona has a particular focus on the global, socio-cultural and political influences on infant and young child feeding practices. She is domain editor for the international journal, *Maternal & Child Nutrition* and has worked on WHO, UNICEF, Government (DH), UK National Institute of Clinical Excellence (NICE), British Council and Commonwealth Secretariat funded projects.

**Naoko Hashimoto** has a Ph.D. in midwifery and is a Research associate at TVU in London, a Japanese midwife and an independent midwifery researcher, her main interests are in breastfeeding and the cultural and philosophical aspects of midwifery practice.

**Christine McCourt** studied for her degree and doctorate in social anthropology at the London School of Economics. Her doctoral work, an ethnographic study of the closure of a long-stay psychiatric hospital, was explicitly intended to be 'applied anthropology'. From 2006 to 2010 she was Professor of Anthropology and Health, at Thames Valley University and she is now Professor of Maternal and Child Health at City University London. Her research and teaching is mainly focused on culture and organisation of biomedicine, maternal and infant care, and social and cultural issues affecting women's health.

**Ólöf Ásta Ólafsdóttir** is a midwife who lives and works in Iceland. Since 1996 she has been the head of midwifery studies at the University of Iceland. In 2006 she completed her Ph.D. work from Thames Valley University in London, looking at the birth stories of Icelandic midwives from the 1950s to the present day, and the chapter is in part based on her thesis. Her research focuses on the relationship between midwives and women and its impact on normal birth, safety and midwives' ways of knowing.

**Trudy Stevens**, after qualifying as a midwife, worked in the Maldives, Ascension Islands and Nepal for ten years. Her work with Traditional Birth Attendants taught her the importance of the social context of birth and of 'mid wifery' so she returned to the UK to read Social Anthropology at the University of Cambridge. Her doctoral study used ethnographic methodology to study professionals' experiences of change in the UK maternity services. She is now a senior lecturer at Anglia Ruskin University.

**Denis Walsh** trained as a midwife in Leicester, UK and has worked in a variety of midwifery environments. He is now Reader in Normal Birth at the University of Central Lancashire and an Independent Midwifery Consultant, teaching on evidence and normal birth across Europe and Australia. He publishes widely on normal birth and has recently had his first book published in the birth centre model.

**Clare Winter** qualified as a Midwife in 1985, then in 1987 practised as an Independent Midwife for 17 years in London. She is now a midwifery lecturer at the University of Dundee and works as a midwife in the Montrose Midwifery Led Unit in Scotland.

# INDEX